# RED DELTA

## FIGHTING FOR LIFE AT THE END OF THE COLORADO RIVER

CHARLES BERGMAN     *Defenders of Wildlife*

FULCRUM PUBLISHING
Golden, Colorado

*For Susan, whose love sustains me*

Library of Congress Cataloging-in-Publication Data

Bergman, Charles.
 Red delta : fighting for life at the end of the
Colorado River / Charles Bergman.
    p. cm.
Includes bibliographical references and index.
 ISBN 1-55591-460-8 (paperback : acid-free paper)
 1. Colorado River Delta (Mexico)—Environmental
conditions. 2. Environmental degradation—
Mexico—Colorado River Delta. I. Title.
GE160.M6 B47 2002
333.91′62′0972—dc21

                         2002003883

Editorial: Marlene Blessing, Daniel Forrest-Bank,
    Julie Van Pelt
Design: Elizabeth Watson
Map (page 9) and graph (page 118): Marge Mueller,
    Gray Mouse Graphics
Cover image: The mouth of the Welton–Mohawk
    Canal teems with birdlife in a red delta sunset.
    Copyright © Charles Bergman.

The poem "Los Ríos Acuden" from the collection
*Canto General* by Pablo Neruda (published by Catedra
in Madrid: 1950, 2000) is used by permission of the
Fundación Pablo Neruda in Barcelona, Spain.

Fulcrum Publishing
16100 Table Mountain Parkway, Suite 300
Golden, Colorado 80403
(800) 992-2908 • (303) 277-1623
www.fulcrum-books.com

Printed in China
0 9 8 7 6 5 4 3 2 1

# Contents

# Acknowledgments

I have been fortunate in this project. The book itself has been a joy both to research and to write. It's impossible for me to adequately convey my gratitude and my debts. To everyone who has helped me on this project, my great thanks.

To the people at Defenders of Wildlife I am especially grateful. I thank Roger DiSilvestro, formerly with the Defenders, for the conversation that led to this book project. Nicole Stoduto and Kara Gillon have been wonderful to work with—great help in research and excellent readers of the early manuscript. My special thanks go to Bill Snape, for his vision, his unstinting support, and his insightful and encouraging reading of the early draft. The delta has an excellent advocate in him. I have enjoyed our relationship immensely.

The research for *Red Delta* began during a Fulbright Senior Scholar Fellowship in Mexico. I owe a particular debt of thanks to the Fulbright-Garcia Robles program in Mexico and to Joan Landeros at Universidad LaSalle in Mexico City.

My debts to all the people with whom I traveled and lived in the delta are implicit in the pages of this book. Still, I want especially to thank the people in the Reserva de la Biósfera Alto Golfo y Delta del Río Colorado in Mexico. Without their superb help and support, not to mention camaraderie and friendship, this book simply would not have been possible. José Campoy is always enormously busy, but he took time to give me tours, share information, and make connections.

José's wife, Martha Román of IMADES (Instituto del Medio Ambiente y Desarrollo Sustentable del Estado de Sonora), helped me find contacts and resources. María Jesús Martínez Contreras showed me friendship and laughed when she found me in camouflage at the Ciénega, waiting for rails, looking like "el monstruo de la Ciénega." Also thanks and *saludos* to Miriam Lara Flores, with her cheerful smile and unflagging interest in the delta's biology.

Ed Glenn went far out of his way on several occasions to help me in my research. Without his aid I can't imagine having written this book. Fred Croxen III of the Bureau of Reclamation was a superb companion in the desert and taught me more than paleontology. And I'll never forget the great experiences and wonderful laughs with Karl Flessa, "surrounded by acres of clams." Thanks also to all in this paragraph for reading parts of the early manuscripts.

Chuck Minckley was an excellent resource on desert freshwater fishes and I thank him for both the information and for reading parts of the manuscript. Michael Cohen of the Pacific Institute also provided an excellent reading.

These researchers helped educate my head about the delta. But many in the delta opened themselves to me and educated my heart. I cannot overstate my thanks to them. Don Onesimo Gonzalez of the Mexican Cucapá tribe figures prominently in *Red Delta*, but he needs to be thanked again here. Additional thanks go to Monica Gonzalez Sainz, his daughter and leader of the tribe, for her gentle soul and deep commitment to her delta.

To the people of the Ejido Luis Encinas Johnson a special thanks. Especially I have to single out Juan Butrón, the great protector of the Ciénega de Santa Clara, and his son, José Juan, for their many hours with me out in the marsh. I came to love this dusty little hamlet of homes near the marsh.

I want also to offer a special thanks to Osvel Hinojosa and Yamilett Carrillo. Also to Jaqueline García-Hernandez. I'm inspired by the work of all of them.

Closer to my own home, I want to thank Duncan Foley for helping me with the geology and for reading parts of the manuscript. Our conversations were a genuine pleasure.

I extend my heartfelt gratitude for the support of Paul Menzel, Provost at Pacific Lutheran University, especially for helping arrange the time away that made the research and writing of *Red Delta* possible. Barbara Temple-Thurston, Dean of Humanities, has offered ongoing institutional support and emotional inspiration. Sue Golden and the people in interlibrary loan, and the library staff generally, were wonderful. Thanks also to Gale Egbers, reference librarian. I also want to thank Maryanne Ashton and Anna Dodge for all the prompt photocopying and for returning books to the library. Tracy Williamson always provided excellent and efficient work and I appreciate her help.

I have long wanted the opportunity to work with Marlene Blessing, my editor on this book. It has been all I had hoped it would be. Daniel Forrest-Bank managed to turn a manuscript into a book with great grace and skill.

To Marilyn Davie, thank you for everything over the years.

Three cats were part of every page I wrote in this book. Sonny rode on my shoulders every morning. Comet sprawled across the desk and guarded the window. Sassy waited patiently for each morning's work. These cats helped me write *Red Delta*.

I need to thank my sons, Ian and Eric, whom I love deeply and who have been more support and inspiration than they can know. I am proud of you both.

Finally, to Susan Mann. Your patience, your insight, and your love have enriched this book. Not to mention my life. This book is for you.

—Charles Bergman

*Defenders of Wildlife gratefully acknowledges Caroline Gabel, whose generous support made this book possible.*

# Foreword

The Colorado River was once an unbroken ribbon of life from the northern Rockies of central Wyoming through the vast arid Southwest into Mexico and eventually the Gulf of California. But today the Colorado River is not healthy. In fact, it no longer even exists in a form that could reasonably be called a river in its southernmost reaches.

By the early 1960s, decades of massive U.S. dam projects and water diversions had reduced the southern flow to a trickle, until not a drop of water reached the Gulf of California. Wildlife was extirpated and the once incredibly lush delta ecosystem dried up. Local fishermen were put out of business. Indigenous people lost a way of life they had led for millennia.

In recent years, however, because of an aberration in the management of the Colorado River in the United States, there has been a small, accidental flow of "surplus" water that has begun to restore the delta's wildlife and its ecosystem, the fishery, and even the indigenous people's way of life. Yet as *Red Delta* explains, this remarkable restoration is seriously at risk. The U.S. government, Colorado River Basin States, and other U.S. governmental entities are hard at work trying to eliminate the accidental flow and redirect it entirely to U.S. uses. This book tells the story of how the current situation came to be and makes a compelling case for revising the binational agreement allocating the river's waters to reduce the massive ecological, economic and cultural harm that has been caused in Mexico.

In some ways, this story of ecological destruction and its consequences is similar to that which could be told about the environmental impact of much of the rapid, don't-worry-about-the-environment rush to maximize industrial development and economic growth in the United States throughout most of the twentieth century, particularly after World War II. But this story has an additional, dark twist, one rooted in what can be recognized today as a grievous example of U.S. economic imperialism.

Ask yourself: If the borders of the United States had encompassed the entire stretch of the Colorado River from Wyoming to the Gulf of California, would the U.S. and state governments have permitted the delta and upper reaches of the gulf to be ecologically devastated? Such an outcome would have been inconceivable.

One of the strengths of this book is that it captures so well (and colorfully) the many sides of the Colorado River equation in real and personal terms. U.S. entities have themselves been fighting over the river for decades, whether it has been one state versus another in the U.S. Supreme Court or the seven Basin States at each other's throats at almost every water meeting convened. To this already flammable mixture have only recently been added scientists, environmentalists and other public interest advocates—although they have been allowed to join in only some of the proceedings, and then as clear second-class participants. Native peoples too are now present, in the form of the Cocopah Indian Nation of southern Arizona, which is struggling with the impacts of the U.S. government's "modern" Colorado River management, and their fellow tribal members (the Cucupa) in Mexico. The Government of Mexico of course is also a player, albeit one that is obviously torn between its desires to fight for what is right for its own lands and people and its needs to placate a powerful and demanding northern neighbor.

After all is said and done, to my mind the over-riding issue is one of cross-border equity in claims on and use of the river. In fact, viewed objectively by modern international standards, one can conclude that the U.S. basically stole the Colorado River from Mexico—and continues to steal it every day it does not allow sufficient water to cross the border to provide for the reasonable ecological, economic and cultural health of the delta and gulf.

Major Colorado River water users in the United States will object to this stark statement. They point to the U.S.–Mexico treaty of 1944, which established the current, grossly disproportionate allocations of river water between the two nations. But there is ample history to demonstrate that the simple existence of a treaty does not assure its fairness, especially when it was negotiated between parties of unequal power.

There are numerous examples of situations in the United States where, notwithstanding the formal legalities of actions taken at an earlier point in time, a current and more enlightened awakening as to what is right has led our government to make up for past inequities. True, these examples have involved primarily internal U.S. issues—for example, belated recognition of the unfair exploitation of numerous Native American tribes and other minorities—but basic issues of right and wrong should not be, and increasingly are not, perverted by the existence of an international border.

We live in an age in which, fortunately, the community of nations has been increasingly and officially subscribing to the view that our planet and its resources are finite. Some problems, including widespread and growing ecological decay, must be viewed in the context of the global commons. Certainly the issue of the appropriation of a major river shared by two nations is in that category.

❦

There does not yet exist sufficient international environmental law to resolve conflicts over the vastly disproportionate claims on the waters of this transcontinental river. But basic principles that guide that evolving law have been clearly enunciated and adopted by the community of nations. International courts and national governments have agreed that every country has not only the sovereign right to manage its own affairs but also the solemn responsibility to ensure that activities within its borders do not cause environmental harm elsewhere. The United States has been a champion of this principle against transboundary harm since its inception early in the twentieth century—ironically even throughout the period in which it was aggressively capturing a river that flowed through its lands into those of another country.

Numerous treaties and international documents to which the United States is a party make ample use of this principle. (For example, the U.S.–ratified 1972 Stockholm Declaration on the Human Environment states explicitly that nations have the clear responsibility to assure that their activities do not cause damage to the environment of other nations.) And well they should, for it is a matter of common sense and simple fairness. I recognize that giving up a claim on valuable natural resources, no matter how unreasonably and unfairly acquired, is not easy. If it were, my organization, Defenders of Wildlife, would not be in court right now—along with other U.S. and Mexican public interest organizations—trying to compel the United States and five southwest state governments to do just that. It is also not clear, given the imperfection of the laws at this point, that this legal action will necessarily produce a result that could be judged fair by the now-established principles of international environmental law.

But such a court result shouldn't be necessary. The U.S. government should take the initiative, as it has in similar situations, to acknowledge the unfairness of the 1944 treaty and negotiate the kind of equitable solution expected of a great nation.

In philosophy it is generally agreed that when we have knowledge of the harmful consequences of our actions, options available, and the capacity to choose those options, then we also have the moral responsibility to act. *Red Delta* clearly describes the serious ecological, economic and cultural damage we are causing to the Colorado River. Scientific analysis indicates that a mere fraction of the current "surplus" accidental flow, provided through perennial and occasional pulse flows, is all that is necessary to sustain the native habitats of the delta and gulf. Negotiating a new "minute," or amendment, to the 1944 treaty is clearly possible if the U.S. government will muster the political will to do so. We have the moral responsibility to act.

—Rodger Schlickeisen
*President, Defenders of Wildlife*

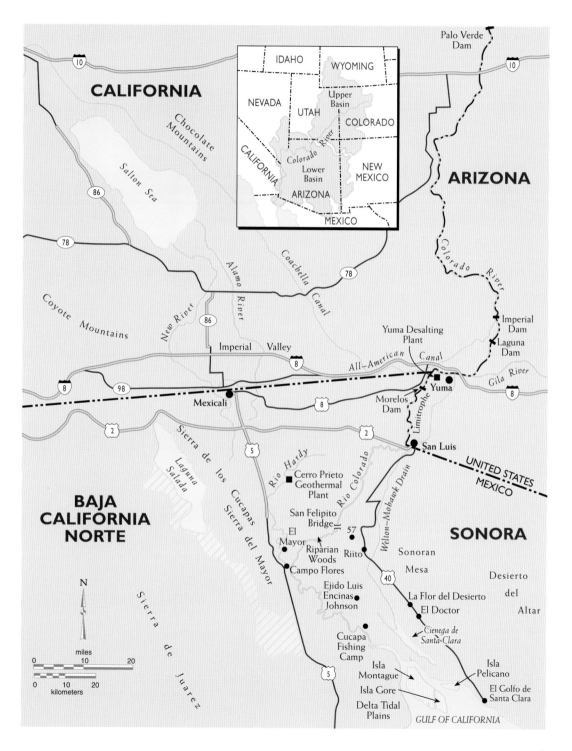

CALIFORNIA

Chocolate
Mountains

Salton Sea

Coyote Mountains

Alamo River

New River

Coachella Canal

Imperial Valley

IDAHO

WYOMING

NEVADA

UTAH

Upper
Basin

COLORADO

CALIFORNIA

Colorado River

Lower
Basin

NEW
MEXICO

ARIZONA

MEXICO

ARIZONA

Colorado River

Palo Verde
Dam

Imperial
Dam

Laguna
Dam

Yuma Desalting
Plant

All-American Canal

Gila River

Yuma

Morelos
Dam

Mexicali

BAJA
CALIFORNIA
NORTE

Sierra de los Cucapas

Sierra del Mayor

Laguna
Salada

Rio Hardy

Cerro Prieto
Geothermal
Plant

San Felipito
Bridge

El
Mayor

Riparian
Woods

Campo Flores

Rio Colorado

Wellton-Mohawk Drain

Limitrophe

San Luis

UNITED STATES
MEXICO

SONORA

Sonoran
Mesa

Desierto
del
Altar

57

Riito

Ejido Luis
Encinas
Johnson

La Flor del Desierto

El Doctor

Cienega de
Santa Clara

Isla
Pelicano

Cucapa
Fishing
Camp

Isla
Montague

Isla Gore

Delta Tidal
Plains

El Golfo de
Santa Clara

GULF OF CALIFORNIA

N

miles
0    10    20

0    10    20
kilometers

Sierra de Juarez

9

# A Brief Note on Language

*Tu . . . me contaste el amanecer*
*de la tierra, la poderosa*
*paz de tu reino . . .*
*y luego te vi entregarle al mar*
*dividido en bocas y senos,*
*ancho y florido, murmurando*
*una historia color de sangre.*

*You sang to me the dawn*
*of the earth, the powerful*
*peace of your kingdom . . .*
*and then I saw you deliver yourself to the sea*
*divided in mouths and breasts,*
*wide and flowery, murmuring*
*a story the color of blood.*

—Pablo Neruda, from "Los Ríos Acuden"
("The Rivers Come Forth") in *Canto General*

One of the intriguing features of writing *Red Delta* is that it concerns itself with a part of the North American continent ruled by two nations, Mexico and the United States. They have different national languages, Spanish and English, and different ways of referring to the same things. I found myself having to make several stylistic choices as I wrote. What am I to call things, especially if the two languages have different names for the same thing?

The question is not whether I should use English or Spanish names, per se, but something more subtle. When the Spanish translation of a place is different from the English name we give it, what should I call it here? I am not referring to animals or plants. With a few conspicuous exceptions, I have used their English names for the sake of clarity and ease of comprehension (though I might introduce the Spanish word).

The question of naming becomes an important issue, however, with regard to people and place names. In *Red Delta* I have made the following choices about language. I have tried to make the choices with as much sensitivity to the ethics of language as I could. Beyond that, I have tried to be consistent.

Following geologists and politicians, I consider both Mexico and the United States part of the continent, North America. I refer to citizens of the United States as "Americans," even though technically all inhabitants of the three Americas—North, Central, and South—are Americans. The designation is less awkward and cumbersome than "North Americans." It is no more precise, either, since Mexicans will refer to themselves sometimes as *norteamericanos*, and sometimes will reserve this for citizens of the United States.

Citizens of Mexico are regularly referred to as Mexicans, unless they are indigenous, as is the case in the Cucapá people. The spelling Cocopah signifies the tribe in the United States. The spelling Cucapá refers to the Mexican portion of this indigenous people.

The large inland sea between the Baja Peninsula and the west coast of Mexico is called both the Sea of Cortés and the Gulf of California. One usually thinks of the Sea of Cortés as the Mexican name for the sea, and the Gulf of California as the American. Yet I have read it referred to variously in each language. In *Red Delta,* I have chosen to use Gulf of California, largely to reflect the language usage in the great biosphere reserve that includes the northern part of the Gulf and much of the lower Mexican delta: Reserva de la Biósfera Alto Golfo de California y Delta del Río Colorado.

When I refer to "the delta," I mean to include all of the delta regardless of the national boundary. When referring to the delta in one or the other country, I will designate it: Mexican delta or American delta.

Finally, I realize that many English speakers will not understand Spanish. Yet the focus of much of *Red Delta* is on the Mexican delta and much of the research was conducted in Spanish—including interviews, in-the-field experiences, and reviews of literature on the delta as a whole. When the Spanish phrasing has seemed important—either particularly colorful or significantly nuanced—I have decided to quote it here. Plus there is a pleasure in the original language that people who know Spanish will, I hope, enjoy. According to convention, the Spanish words and phrases are italicized; direct quotations are not. A translation follows immediately, except where the meaning seems so apparent as not to require translation. Unless otherwise specified, all translations are mine.

# One Delta Live

*Our little sloop floated down the Red River on our way to
the Vermillion Sea* [sic]*, on a February afternoon in 1904. . . .*

—D. T. MacDougal
"A Voyage below Sea Level on the Salton Sea"

THE SMALL PLANE looks like a bee, painted brown and yellow. The paint is fading, but the single engine is strong. It's a workhorse plane, a 1956 Cessna 182, that Sandy Lanham's been flying for years. She is fifty-three—"a good age," she says—has brown hair and eyes, and prefers a mirror with a light coat of dust. It softens the focus, she says.

She calls herself an environmental pilot and describes what she does as "flying to protect wildlife and save ground—including the sea." It means she has dedicated herself to helping researchers and environmentalists study wild regions along the border in the arid Southwest and throughout Mexico.

Because she is enormously skilled and provides her services at reasonable rates—flying can be very expensive—Sandy is famous among researchers and conservationists in the region. She doesn't fly to get rich. She lives in a straw-bale house in a canyon outside Tucson, which she says "boosts her soul" and gives her reasons to feel grateful. She does it because "the work matters." She raises the money and then makes research flights. Through her surveys she helps protect shorebirds and jaguars, pronghorn antelope and clams, blue whales and blue crabs. "All things big and small," she says.

Recently the quality and value of her life of work, studying and protecting the borderlands and Mexico, has been recognized with a MacArthur Fellowship, a "genius grant," which will help support her work.

◄ *Wispy pink clouds
herald the morning
over the Ciénega de
Santa Clara, the
world-class marsh
that has made the
delta a conservation
priority in both
Mexico and the
United States.*

▲ *A group of white
pelicans wade in the
shallow waters of
the Ciénega.*

Sandy has been intimately involved in the new effort to study and restore the delta of the Colorado River in Mexico. She has helped researchers create a new map and a new attitude for one of the most neglected and abused landscapes in North America.

She and I have been flying over the delta for a couple of hours already. She rolls the plane deftly into a tight turn. It bellies heavily into the sun-yellow morning air, and we begin a thrilling spiral upward into the sky. She wants to give me a wider view of the delta, which lies below us like a vast bowl between the Sonoran Desert to the east and the mountains on the Baja Peninsula to the west. In the south, the head of the Gulf of California gleams in the reflected desert sun, and hundreds of thousands of acres of farmlands stretch northward from Mexico into California's Imperial Valley.

Then Sandy makes an almost off-handed comment. It's a nudge that pushes me into a new view of one of the saddest river deltas anywhere in the United States or Mexico.

"It's a devastated delta," Sandy says through the intercom. "But," she says, "I love it."

With those words a new set of feelings suddenly popped into clear focus. They had been taking shape for nearly two years, as I conducted my research throughout the delta. With her words, though, I realized that I too had come to love this delta. Among other things, this book is the story of how and why I would come to love this place—and want to save it.

That people have slowly come to care about the Mexican delta of the Colorado River is one of the most remarkable environmental stories on the continent. Researchers like Sandy Lanham, and many more whose stories are told in this book, are coming to the delta after it was ignored for virtually all of the last century. During that time, the delta became one of the major symbols in two nations of an environmental disaster area.

Yet this new generation of researchers has brought new eyes to the delta and the place has been "rediscovered." What they have found in the "devastated" delta is an almost miraculous turnaround, an ecological revival that is as wonderful as it is improbable. The revival is wholly accidental, the result of stunning bureaucratic mistakes and water-management decisions. It defies the national projects in both the United States and Mexico that would subject the complete length of the Colorado River to a program of exhaustive total use. Sadly, it may be temporary.

Out of this recovery and rediscovery, the delta has emerged as one of the top conservation priorities of environmentalists in both the United States and Mexico. It has also become arguably one of the single most important binational environmental challenges on the continent. We now realize the delta was probably the richest biological area in the southwestern part of the North America.

The debate over the delta—if it should be saved and how—has placed it smack in the middle of a new and intense round of water wars in the United States. Moreover, because the delta contains spectacular areas of crucial biological value, it has become a compelling force in changing the terms of these water wars.

The Mexican delta of the Colorado River, through its dramatic recovery, is helping finally to endow this ecosystem with genuine status in the legal debates about how the water of the Colorado River is managed. Few accomplishments in the water-strapped West are more important and more difficult.

*Red Delta* uses experiences with people in the region—like this one with Sandy Lanham in her beloved Cessna—to tell a double story: first, it tells of the improbable and wonderful regeneration of the delta in Mexico, however fragile; second, it follows the legal and political maneuvers on the part of environmental organizations to save the delta.

As people finally learn to see this place appreciatively, they have come to care about it, moving a long-neglected and terribly insulted part of the continent into international prominence.

We have long known that the delta nearly died, choked and starved for water by a frenzied program of dam building on the Colorado River. The water from the river was diverted by lawyers and engineers in a massive project of nation building in the deserts of the American Southwest. Mexico followed close behind the United States as both rushed to put the water of the river to human use. Any drop of water left in the river and reaching the Gulf of California was widely viewed as wasted. It still is by many people. One of the fastest growing cities in Mexico—Mexicali—is located in the middle of the delta. The delta is also a relatively short drive from some of the fastest growing cities in the United States: Las Vegas and Los Angeles, San Diego and Phoenix. All of these urban centers were created or are sustained in this desert by water that once flowed almost the width of the nation and into the delta of the Colorado. Over the course of the last sixty-five years, about 99 percent of the water from the river has been removed and redirected for farms and cities. In a fundamental way, we stole the Colorado River not just from Mexico but from its delta, turning the river against itself.

This book does not treat the delta, however, as a symbol of environmental devastation. Nor is this book a critique of unchecked American and Mexican growth in the deserts. Most especially, this book is not an elegy to a lost place, as so much environmental literature has been in the past.

What this book is about, is a positive story of recovery—and of the fight to preserve what has been accidentally restored to us. The delta is part of a new stage in environmental efforts in the West—devoted to restoration. The delta as an issue is less about opposing dams than about restoring habitats and communities. It is at the heart of a major effort to define new practices for living sustainably with water and the environment.

The delta in Mexico is the focus of this book precisely because it offers some of the best restoration and conservation opportunities in the entire continent. This is contrary to all expectation and puts the lower Colorado River in the United States to shame. The delta underscores the unity of the river and its delta. It is alive and made whole by both natural and human history. Decisions made about Colorado River water in the United States affect this superb and regenerating ecosystem—yet the delta in Mexico and its endangered creatures still have no protection under U.S. law. Thus, U.S. water managers have no mandate to restore the delta.

It is one delta live. To preserve its vitality, we need to work out new models of international environmental management.

The book begins with the reconstruction in our imaginations of the delta as a place. And while it was disregarded and abused for most of the past century, it turns out to have been a place that was wonderfully populated—populated by communities of plants, communities of animals, and communities of people. All of these have been left out of our vision of the place and our disposition of the water of the river.

Through a series of narratives on the landscapes and various ecosystems of the delta in Mexico, the first half of the book offers a new vision to us of this place—a startling new vision that contradicts virtually every cliché and stereotype Americans are likely to hold about the place. The first half of the book offers the reader "eyes," with an imaginative re-creation of the delta based on the discoveries of researchers and local people living there. In the process, it slowly assembles a new map, a map that includes human communities that have been largely invisible, as well as vibrant plant and animal communities that few people would have imagined could thrive just below the international border.

Such a project of re-mapping and re-imagining, as it has been conducted by researchers, has been revolutionary in changing our understanding of the delta. It has opened up the landscape to new vision and new possibilities. Since Hoover Dam went on line on the Colorado River in 1935, the delta of the Colorado River has been a blind spot in the American imagination. Americans took the river's water and turned their backs on the portion of the delta below the border.

One of the major studies of the Colorado River is *A River No More: The Colorado River and the West*, by Philip L. Fradkin Jr. It was first published in 1968 and has been expanded and updated into the 1990s. The Bureau of Reclamation largely wrote the new map of the river: they controlled the river, they defined the human geography of the West. Fradkin shows how the delta was cut off from the river in the process, noting that "Almost all the maps drawn by the federal water agency and used by the western states show nothing, or at most, the lone detail of the river flowing the last few miles through the featureless foreign land." (Fradkin [1968] 1996, 292)

It is as if, he goes on, "Mexico did not exist as a viable country." For Americans the delta was a kind of vast uninhabited wasteland. Empty. A terra incognita. Out of sight and out of mind.

Even for Mexicans, the only "real" parts of the delta came to be the ones receiving water—farmlands and the growing cities. The delta below the levees— the unfarmed delta—became a blind spot to the Mexicans as well.

The delta has always challenged westerners. It is one of the most difficult and daunting landscapes on the entire continent. It occupies the trough created by the great fault running from the Gulf of California up to the mountain pass north of the Salton Sea—an area nearly the size of the state of Connecticut. It's the hottest place in North America. Temperatures in the summer regularly climb past 120 degrees Fahrenheit. It gets only a couple of inches of rain per year. The delta was, ironically, one of the first places in North America that Spaniards explored. But it has remained one of the least known.

Less than 5 percent of this delta remains wild. Most of that is in Mexico.

And this blank spot of the Mexican delta on the Bureau of Reclamation's maps is a new version of this "invisible delta." This was a willful invisibility for the region after Americans built the dams and took possession of the delta's river. It has been convenient for Americans to ignore the delta—to choose not to know what is happening there. If the delta were a "zero to the south," we need not think about it or concern ourselves with the environmental consequences of robbing the delta of its river.

After the dams, the delta was abandoned. We didn't want to stop to tally the costs rising in another country. Besides, it was already too late. The delta was dead and gone. A dead delta? Then no one has to change their behavior—including the Bureau of Reclamation and the water users in the United States and Mexico—for environmental restoration in the delta. You can't conserve what is already lost.

Yet the delta survived, even began to thrive, in spite of the nation-building programs to the north. A few researchers actually began to visit the delta, and they had to learn to see the place all over again.

If you want to find images of a wasteland, you still can in the delta. At times in the summer, huge winds kick up dust storms that turn the midday dark as night. They are called *viento negro*, black wind. There is crushing human poverty and ecological poverty as well. You will see salted-out landscapes and huge desolate mudflats. But if you look beyond the obvious, or if you let local people teach you, you will begin to discover another delta. A new delta. This new delta has gradually emerged front and center in our awareness, and has transformed our view of the place and our view of environmental possibilities for it.

When biologists and geologists began flying over the delta with Sandy Lanham, they realized there were newly created habitats in the region. Most were created by "accident," the results of bureaucratic mistakes and inadvertent decisions by United States water managers upriver. The amazing stories of these "mistakes" and their unanticipated ecological benefits in creating this new delta—the delta that is proving so troublesome to the very agencies that unintentionally created them—inform much of the narratives in Part I.

You could call this new delta an accidental landscape.

*The salty Ciénega de Santa Clara—the Ciénega for short— is one of the greatest marshes on the North American continent.*

Thanks to miraculous human blunders and very heavy rains, the delta sprouted some world-class wetlands and utterly unexpected forests of cottonwoods and willows. We are still learning what is there—seeing it really for the first time. Much of what is in the "new" delta is still not known, still not documented. What is known, however, is that the new forests and marshes of the delta support a level of wildlife that hints at the landscape's former abundance. Several of the premier endangered species of birds and fish in the United States now rely on the Mexican portion of the delta—including Yuma clapper rails,

desert pupfish, and southwestern willow flycatchers. The delta sustains populations of these and other species—populations that are almost certainly critical to the survival of these species, even in the United States. The Mexican portion of the delta has become, in effect, a recovery zone for many of these species.

Populations of some endangered species in the Mexican delta are found nowhere else in the world. This is true of the endangered vaquita, the world's smallest porpoise. It was discovered only in the 1950s and remains the rarest, most endangered marine mammal in the world, its entire range limited to the upper Gulf. Totoaba, also found in the upper Gulf, was once a hugely important commercial fish and is now nearly extinct. The delta has also become a critical link in the Pacific Flyway for migratory birds.

The narratives of Part I describe these species and habitats, along with others—including human communities. Many Mexican communities have discovered that they have ecological connections to the land they had not before understood, and a new ecological consciousness is emerging as they fight for rights that they have long been denied. The narratives also place the new delta in the context of its geological history and the strangely paradoxical state in which it now finds itself. The post-dam delta is, as geologists say, an "anti-delta."

The signs of life in this accidental landscape, all supported by unintended flows of water, have inspired a new interest in preserving the delta. The initiatives have been diverse, but they share a single goal: to find legal guarantees to safeguard the accidental recovery.

Through local community projects, on-the-ground environmental restoration efforts, and initiatives at the highest levels of government, a wide range of people are now at work on behalf of the Mexican delta. Even the United States Secretary of the Interior and the Mexican Secretary of the Environment have entered into negotiations and discussions, resulting in joint statements on the importance of the delta to both countries. Hands have reached across the boundaries in binational initiatives. La problematica del delta—the problem of the delta—has become one of the most ambitious and multifaceted environmental initiatives yet between the citizens and governments of the two nations.

The problem of the delta can be summed up in the single most important word in the arid West: water. The questions are how to get it and where should it come from?

This dilemma is the central subject of Part II of Red Delta.

The inevitable answer, the controversial answer, the incendiary answer is obvious: the United States. The waters that have created the revival in the delta all came from the United States and originated from water management decisions there. By current law the United States controls 90 percent of the allocated water in the Colorado River. The United States built the dams on the river that diverted the water, which led directly to the devastation on the delta.

It is time, say environmentalists, that the United States help to restore an ecosystem it has done so much to ruin, even as it has benefited so materially from the delta's destruction. It's a convincing argument, since the allocation of the river's water by law has not been equitable. Many communities—human and nonhuman—have been left out of the codified dispensations. Perhaps the most compelling and poignant story is that of the Mexican Cucapá Indians. They are "the people of the river," yet they have been completely deprived of any river water. The river people now live in a sad village without water. In danger of going extinct as a tribe, theirs is one of the most crushing examples of environmental injustice to be found anywhere. It hits very close to home.

In the West, water is life. For some it is even more important. It is money and power. A complex body of law, collectively the "law of the river," controls its allocation. The Colorado River perhaps is the most heavily regulated and heavily litigated river in the world. It is this law of the river that controls how the river is regulated, how its waters are divided. Through the struggles to assure guarantees of water for the Mexican delta for ecological restoration, the delta has been thrust into the midst of some of the most contentious and vicious infighting imaginable.

Interests that benefit from the current legal disposition of the river's waters in general want no part of these reallocations of water. They don't let go of their water rights easily. A new round of water wars is under way in the already over-populated and still-growing West. Making cross-border deliveries of United States' water has led to intense resistance from the states, irrigation districts, cities, and other users—the "water buffaloes," as they are often called, which are the main beneficiaries of the current arrangement of water deals on the river.

The body of law—the law of the river—is almost sacred to many in the West. It's right up there with the Ten Commandments. Yet environmentalists hope they are close to making groundbreaking changes in this law of the river. The significance of these environmental and international changes is impossible to overstate, both in their relationship to nature and to Mexico.

Environmentalists have focused on two basic strategies to change the law of the river. The first involves bringing a suit that uses an innovative application of the Endangered Species Act. The species in the Mexican delta impacted by the suit are on the U.S. list of threatened and endangered species. Lawyers for a coalition of groups led by Defenders of Wildlife argue that there is nothing in the Endangered Species Act that limits its application to the United States alone. It requires federal agencies to protect and help endangered species harmed by their actions and makes no exemptions just because the species happen to be found in another country. In other words, operating dams in the United States may harm species in Mexico, in which case the Bureau of Reclamation would be required to make positive efforts to help, or "mitigate," those negative impacts.

The second major strategy pursues nothing less than an amendment to the 1944 United States–Mexico Water Treaty that established the legal division of water between the two countries. The treaty was only signed after nearly a century of often-bitter negotiating between the two countries and has been tantamount to a liquid border between them. The issues of how to acknowledge a "one live delta" and to support it with a continuing flow of water are not yet resolved. The delta hangs in the balance, and much depends on Mexico's resolve to move dramatically forward in its own environmental commitments on behalf of the delta.

Beneath the legal issues, the restoration of the Mexican delta of the Colorado River raises questions that are fundamental to our relationship to water and nature as a society. What would a new water ethics involve and how would it change the way we treat water as a commodity? How can we develop a way of managing the river that is more sustained and more equitable, even as water supplies are increasingly over-tapped? How can we support a concern for ecological restoration in the face of the relentless, almost senseless population growth in the West? What national security issues are raised by an unfair allocation of water between two nations, one powerful and controlling the water, the other developing and water poor? How have social values changed in the West and how are those changes driving legal changes in the status of nature?

One of the great ironies about the restoration of the Mexican delta is that this neglected part of the river turns out to have the largest area left for restoration. Yet this delta is completely hemmed in by human development. This large "wilderness" is completely artificial. Even the revivals of the delta have been artificial, in that they derive entirely from human decisions involving water upriver. The delta hints at a new, an almost postmodern balance between nature and culture in the West.

The law of the river, for example, is more than the legal requirement for securing water for the delta. The law of the river illustrates how thoroughly mediated our relationship to water has become. This body of law has theoretically allocated every drop of the Colorado River. It is a separate river, a textual river, a river of words that stretches back over a century in time. First we created a language that defined and controlled the river, and then that language transformed the river into a simulacrum of itself, a vast plumbing system.

The law of the river gives us a virtual river, a second river—and this river controls our access to and understanding of the real river. Such an altered consciousness is not a uniquely American phenomenon, but Americans have certainly perfected it—the preference for the virtual reality over the physical reality. The virtual has become real. To change the actual river, we have to change this virtual river.

More fundamentally, to change the relation of the delta to its own river, we have to rewrite the river of our imaginations. At the deepest level, this is the goal of *Red Delta*: to reconnect the delta to the river and reimagine it as a living place, a place rich in human and natural life, a place rich in meaning and value.

Many people believe that the time of the delta has finally come. The obstacles are real and very serious. Yet the visibility of the new delta is creating a new international celebrity status for this place, at least slowly, as it moves to the top of people's to-save lists. It has made it onto the environmental radar screen.

In 1998, for example, the environmental organization American Rivers listed the delta of the Colorado River as one of America's ten most endangered rivers. In reality the delta has been endangered for nearly a century. What the listing shows is that people are beginning to care enough to want to see the delta protected and restored.

*Red Delta* is about the why and the how of that caring.

The title of this book, *Red Delta*, honors both the endangered status of the delta and the desire to restore it to some semblance of its former glory. The Colorado River takes its name from the spectacular silts it once carried before the dams, soils that turned the river a unique shade of red. The name *colorado* means "red" in Spanish. Even the Gulf of California was originally called the *Mar Bermejo* by the first Spanish explorers in the area—the "Vermilion Sea." The title recognizes the original river and the fertile soils out of which it created its own rich delta. Red is the color that was once intimately associated with the whole watershed system of river, delta, and sea.

Red is also the color of threat and danger—code red.

The love Sandy Lanham feels for the delta expresses the change in attitude toward the delta in recent years. It's an example of seeing the richness of and caring for this overlooked and endangered place. As Sandy and I soar high above the delta, we look down upon hundreds of thousands of acres of farmland. We see thousands of acres of brown-out mudflats, salted over and barren. But we also see the new green habitats that have so renewed the delta.

Sandy takes us into a slow dive over one of these green places. We are skimming barely one hundred feet over one of the greatest marshes on the North American continent. It's called the Ciénega de Santa Clara and figures prominently in *Red Delta*. Deep in the marsh cattails, invisible except from the air, a flock of about a hundred fifty white pelicans rests in shallow waters.

Sandy circles skillfully to bring us back around and over the sitting birds. The rippled water flashes in the low angle of the morning sunlight. I photograph the pelicans as we swoop over them. Sandy is expert at this, knowing just how to bring us in for the best shots without disturbing the birds.

In a single day these birds will fly from their roosting spot in this Mexican marsh up to a place in the United States portion of the lower Colorado River. Like so much else in the delta, they are living links between the two sides of the international border, migratory birds that can help make the delta whole.

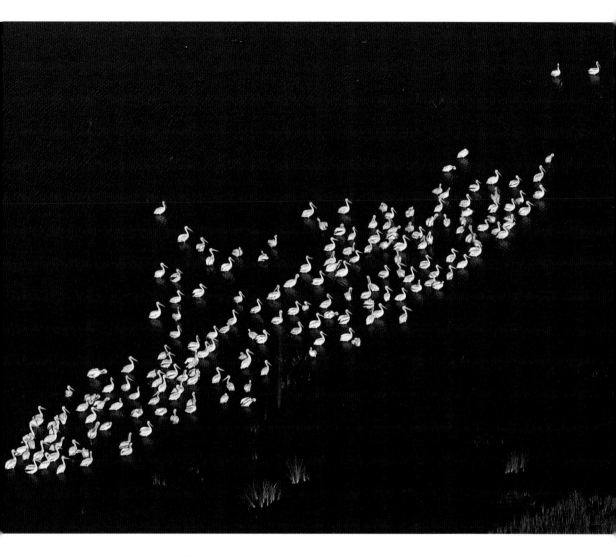

Sandy swings the small Cessna back around for another pass over the white birds. The pelicans come between the plane and the sun. The light of the water flashes again, a burst of light like a mirror exploding. The white birds vanish in the light. It's an illuminated moment—one of those instants where I feel lifted out of myself, almost transcending myself in beauty and joy.

We complete the turn over the pelicans, emerging from the reflected light. Sandy and I are both beaming. Moments like these are why she flies. Why she lives the life she has chosen. Why she loves the delta. This is the delta I care about, a red delta that is full of shining lives, human and creaturely. This is the delta we are inspired to save.

*A flock of white pelicans rests in the shallow waters of the Ciénega. These birds migrate between the United States and Mexico, symbols of this "one delta live."*

# PART I

# Natural History and Human Values

*We can be ethical only in relation to something we can see, feel,*
*understand, love, or otherwise have faith in.*

—Aldo Leopold, "The Land Ethic," in *A Sand County Almanac*

ONE

# The Delta of Secret Lives

ENDANGERED SPECIES: Yuma clapper rail
PLACE: Welton–Mohawk Canal

AS THE ALMOST insufferable heat of the afternoon sags slowly into a comfortable evening warmth, we head to work. We are hurtling down a back road along a canal lined in concrete, kicking up gravel and a long, trailing storm of dust. On either side of the canal, the delta desert looks like a toxic wasteland of saline ponds and salt-frosted, sterile brown dirt. Several miles down the road, the concrete in the canal abruptly ends. The drainage ditch through the desert turns suddenly, surprisingly, into a "natural" flow. Trees and brush and cattails jam the banks. Ahead, the canal bends and disappears into the dense green wall of the marsh.

There is nothing "natural" about this place. The canal itself is really a drainage ditch running thirty-five miles south of the international border at San Luis Río Colorado. It comes from the Welton–Mohawk Irrigation and Drainage District in Arizona. It has a variety of names and is known locally as the Welton–Mohawk Canal. The water in this ditch is considered "waste" water—it's being dumped in Mexico because it is too saline for agriculture and at one time there was nowhere else to get rid of it. The concrete canal actually goes nowhere. It simply ends in the middle of this vast delta plain, as if it were abandoned.

From all appearances, the canal is widely unappreciated, too, suffering more than its share of neglect and abuse. People toss everything into the canal: Nike tennis shoes, plastic milk cartons, beer bottles. Dead animals—cows, goats, horses, dogs—have been discarded here. Humans, too. Earlier this year, two human bodies—one a suspected drug dealer—were found floating in the ditch.

◄ ◄ *Amid a barren sea of salted mud, the Ciénega is a dreamy green oasis, the greatest haven for wildlife in both Sonora and the lower Colorado River. This view looks northeast.*

◄ *The Yuma clapper rail is one of the most severely endangered species in the arid Southwest. The Ciénega is a refuge for the largest population in the world of this species.*

▲ *Rosy light silhouettes a songbird amid the reeds on a soft Ciénega evening.*

27

*The Welton–Mohawk Canal carries United States' "waste water" into Mexico, where it accidentally created this wet desert paradise. Looking for clapper rails, we canoed from the end of the concrete canal, visible in this photo, out into the marsh.*

This does not seem like a place to come for a spectacular natural experience. Yet the waters of this "fake river," if you will, have given birth to one of the most superb and least-known wild places in North America.

José Juan Butrón and I lower a canoe down the bank. José Juan's nickname is Katán, which is what I call him. We hold the canoe next to the bank. I climb in and take the front seat. Katán grabs the equipment we need—tape recorder, hand-held loud speakers, and notebooks—and clambers into the stern. The rest of the team watches as we shove off and slide with the stream.

We are part of a team conducting a census on one of the delta's most wonderful secret lives—the Yuma clapper rail. It's one of the preeminent and most elusive of endangered species in this arid part of North America. Our goal is not just to find a rail. We want to see and photograph one.

The Yuma clapper rail (*Rallus longirostris yumanensis*) is endangered, but it's not one of the celebrated species. You won't find the Yuma clapper rail in the pantheon of famous endangered species that includes bald eagles, Florida panthers, and California condors. You may not even know that it's a bird, a mud-loving, chicken-like marsh bird. It lives a very secretive life, hidden in the

dense marshes of the few wetlands in this desert part of the continent. If few people even know about the undervalued Yuma clapper rail, even fewer yet have seen one.

But in the water-strapped West, the rail may be on its way to stardom, as the delta emerges into the political spotlight of water politics. As an endangered subspecies of a marsh bird in the desert, the Yuma clapper rail lives its life in the middle of the Byzantine water wars along the Colorado River. It is increasingly coming out of hiding and into full view, a representative of all the many endangered and extinct species of animals and birds that once thrived along the lower Colorado River.

Most people would be hard pressed to name any of the endangered species along the lower Colorado River—yet the region has one of the highest rates of extinction and endangerment on the continent. Arizona has about 107 endangered and threatened species, putting it up there with Hawaii as the leader in the United States. The lower Colorado flows through a thirsty world in the western deserts, and people suck it nearly dry at the expense of such unheralded species as the Yuma clapper rail.

The director of the biosphere reserve that has been created by Mexico in the Upper Gulf of California is José Campoy, formerly a biologist specializing in endangered freshwater fish. José estimates that there are about fifty endangered and threatened species in this delta region of Mexico. Half of them, he says, occur on both sides of the border. It's an estimate, because one of the defining features of the delta is that we simply have not begun to study its natural history carefully until very recently.

Beyond being an endangered species, the Yuma clapper rail symbolizes this unknown delta. In the past several years, studies like the one I am helping with have transformed our understanding of the delta and its natural wealth. This new knowledge, in turn, has spurred a growing resolve, finally, to protect and restore the long-neglected delta.

There are eight species in the Rallidae family of birds. Though most are intensely secretive, one of the species is anything but: the ubiquitous and noisy American coot, which looks vaguely like a duck. Most of the rails, though, keep to cover. The black rail—known to be in the delta—and the yellow rail are two of the most difficult birds to find and see in North America. The Yuma clapper rail, the Virginia rail, and the sora rail are all elusive birds that can be found in the delta, creatures of marshlands, backwaters, and sloughs.

Only rarely do the shy Yuma clapper rails come into plain view. When they do, they can be surprisingly bold, sometimes walking right up to you. Because rails are so difficult to see, birdwatchers especially prize them, cherishing every sighting. Several of these species are among the most sought-after species of birds in North America for birders.

It had long been known that Yuma clapper rails occur south of the border as well. But no one had conducted a thorough survey of the birds until Osvel Hinojosa, who leads the research team I have joined.

Osvel is a young Mexican graduate student doing his master's thesis at the University of Arizona on Yuma clapper rails in the Mexican delta of the Colorado River. He is one of the most engaging biologists I have been with in the field, a pleasure to know. He is short, with baby-faced features and long hair to the middle of his back. Easygoing and unflustered by any of the difficulties of working in the delta, he has a lot of people who live in the delta working with him on his current survey. Serious studies of these rails only began in the 1980s. Osvel's study in 2000 would produce the most definitive population estimate yet of this rail in the delta.

Osvel stands on the bank of the canal as Katán and I shove off into the current. We wave goodbye as he heads off to another spot to look for rails. For more than a week, we have been counting birds in various locations in the Ciénega de Santa Clara, the huge cattail marsh in the delta fed by the waters of the Welton–Mohawk Canal where we had launched our canoe—but I have only seen one rail briefly and from a distance. A photograph was out of the question.

Katán and I have chosen to descend this canal on a hunch.

"Es un lugar especial," Katán says. "Casi siempre salen los palmoteadores." It's a special place. Clapper rails almost always come out into view here.

When Katán has a hunch, I pay attention. He was born not far from here, in an *ejido*, a small cooperative farming community. At twenty-five, he has spent his entire life growing up by the marsh created by this canal. He has a crew cut, a very big smile, and a personality that is utterly likeable. He doesn't have much material wealth. With his wife, Alma Rosa, and three young children, he lives in the local *ejido* in a single-room house about fifteen feet square. Katán tells me he earns about four hundred pesos per month at a Korean *maquiladora*, or assembly plant, in the nearby town of San Luis Río Colorado. That's about forty dollars.

As do others in the *ejido*, Katán supplements his income with occasional work, whenever possible in the marsh, as in this participation in Osvel's research. He counts this marsh as part of his heritage and inheritance—part of his wealth. Katán knows the Ciénega intimately. He was born beside it, grew up beside it, and has spent his life exploring it. It is the landscape of his life.

For the first half mile, the canal is about ten yards wide and easy going. Then the tules and cattails crowd into us, the channel narrows, and we are barely able to paddle. Sitting in front, I grab the tules that choke the channel and pull us through, ducking my head while being slapped by the reeds.

The evening sun lowers and the light thickens in the dense tules of the canal. Commenting on the sunset, Katán tells me we have lost "la cobija de los pobres." The sun, he says, is the blanket of the poor.

Stopping occasionally, we play the call of the clapper rail on the tape recorder, aiming the speaker toward the tules. We get no response. For the better part of a week we have been conducting the same drill over the marsh to amazing results. Yuma clapper rails make a distinctive call, a call that gives them their name. The sound is loud, completely unmusical, even comical, a rapid *clack-clack-clack-clack* sound, slowly winding down. It sounds vaguely like a person clapping hands. In Spanish, it's called a *palmoteador*—someone putting palms together. A clapper.

Once clapper rails were locally called "marsh hens," which recognizes their resemblance to chickens. They walk like chickens, with the same jolting motions, and they have a meaty body. Unlike chickens, though, they don't have long tail feathers. Plus, they skulk about on very long legs and have a long beak for mud poking. With their long toes, they're well adapted for walking on lily pads or mud in marshes.

Chickens cluck peacefully. Not marsh hens. They are loud, even a bit frantic. You can hear them all over the marsh. Sometimes they will suddenly let out their wild call from a hidden spot very near to you in the canoe. It's startling every time.

*José Juan Butrón (nicknamed Katán) smiles in the desert light before his home in Ejido Luis Encinas Johnson.*

31

Often we would play a tape in the marshy area and have four or five Yuma clapper rails answer at once. Patiently, quietly, we would sit in the canoe, playing the tape and hoping to get a glimpse of the rail. We could tell they were close, within a foot or two of us, hidden in the reeds. They would clack and clatter tantalizingly, almost as if mocking us. But they would not come out to view.

"Hay muchos pero no salen," as Katán would put it, smiling. There are a lot of 'em, but they won't come out.

That's why Katán suggested this canal. The canal narrows and twists until it reaches the spot where it pours out into the big marsh, creating a fan of muck and fertile mud—a delta within the delta. Just up the canal from this mouth, the canal is about five feet deep in places. But in the bends, it sweeps wide and shallow, creating little openings in the dense tules.

Katán calls them *orillas*, or open flats. Water slides over them in shallow sheets of clear water. The muck is gooey, greenish, and full of small crustaceans and insects and other organisms that birds love to eat. This rich organic muck has its own word in Mexican Spanish—*fango*.

This is the stretch of water we are now negotiating.

We pass three of these clearings and enter a fourth, the widest clearing yet. Katán angles the canoe into the mudflat at the base of a small tamarisk. We are completely surrounded by tules ten feet high, a thick screen of waving green stalks. The open water of the marsh itself is very close, only about one hundred yards down canal. We can't see it through the cattails, but we hear it. Ducks quack. Black-necked stilts scold each other. A green heron zooms low over our heads, scouting for a fishing spot.

Katán plays the tape of the clacking rails. Once, twice, three times. No answer.

I turn to look at him in the canoe and shrug. As I do, Katán's face brightens with excitement. Barely moving, he gestures with his face for me to turn around.

I turn slowly. A Yuma clapper rail has stepped out of the reeds and is standing, completely visible just on the other side of the tamarisk.

Katán and I sit silent and still in the canoe, low to the water, almost eye-to-eye with the fifteen-inch tall bird. We seem to be facing the bird on its turf and on its terms. At its level.

We're enormously excited, but hold our feelings in. The rail, on the other hand, treats us with a casual disregard. It's not so much fearless as unconcerned, engaged completely in its own life. My first thought when I see it is, yes, "chicken."

There is something at once vulnerable and endearing about this group of marsh hens. If they were easier to see and photograph, they might be more widely loved.

Clapper rails are a seashore group of rails, up and down both coasts. The Yuma clapper has the distinction of being a seashore bird found in North America's greatest desert. It manages this feat by having followed the marshes up the Colorado River, giving it the additional distinction of being the only clapper rail to live in nearly fresh, as opposed to salt, waters. Once it was probably abundant in the many marshes and sloughs stretching up the Colorado River into the desert all the way past Yuma. But as the river was dammed and the wetlands lost, its numbers crashed.

The bird walks on its long orange legs in a deliberate, herky-jerky style, chicken-stepping across the mud. Its tail is short, highlighted with white. With each step, its tail twitches and flicks like that of a barnyard bird.

With its long neck and beak and its stilt-like legs, it can strike a surprisingly elegant pose. Its plumage is appropriately muted for a shy bird, with soft tones of grayish brown on its back and pale cinnamon about the throat and chest.

The bird walks in front of us, little jolts of white flashing as it twitch-steps. It stops, looking back at us over its shoulder, and seems to strike a pose. It holds its pose long enough for me to take photographs. Like all rails, its most distinctive feature is its rear end. For reasons that I don't know, rails have lateral stripes across the rump or butt, what in the language of bird topography is called the "vent."

On this rail, the pattern is ripples of black and white and gray, vaguely like the pattern of light off waves of water. You might not at first be inclined to call this rail beautiful. But after spotting the stripes on its butt, you will likely be won over. Rails have arguably the most beautiful butts in the bird world.

The rail then marches right up to the canoe and struts along its entire length. The bird is almost within arm's reach as we sit in amazement like stones. We don't move. We don't even breathe. Neither of us is sure what the rail will do next.

In the course of this two-week survey of the various wetlands in the delta, Osvel Hinojosa produced a population number for Yuma clapper rails that is part of the changing image of the delta of the Colorado River. In the Mexican portion of the delta, Osvel's census determined that there are about 6,300 Yuma clapper rails. By far the largest number of them lives in this marsh, the Ciénega. It's their refuge. This marsh, it turns out, supports the largest population of this endangered species in the entire world. Surveys in the United States produce a population estimate of 500 to 1,000 of these birds, only a fraction of those found in the Mexican delta. In the geography of this endangered species, Mexico in general—and this marsh in particular—is ground zero, absolutely crucial.

It is a result that has surprised many scientists and researchers. Beyond that, the population of endangered rails illustrates that there is a much more abundant and vibrant environment in the delta below the border than anyone had dreamed of, a secret and unknown world of life.

Osvel's study of the Yuma clapper rail and its astonishing abundance in the delta is one of a number of studies that, in the last ten or fifteen years, have transformed our sense of the geography and biology of the delta in Mexico. At the same time that many were dismissing the delta as a symbol of biotic death on a massive scale, Yuma clapper rails and other endangered species were proliferating in these unsuspected habitats. As the region continues to struggle with large-scale degradation from poor water and no water, birds like these have somehow begun to prosper.

The rails also illustrate how connected the ecosystems on both sides of the boundary really are. Consider this. The Ciénega marsh is home to a staggeringly large population of this endangered waterbird. These are in a sense "Mexican birds." Yet they have thrived because of the marsh—a marsh that is created and sustained entirely by water in Mexico that technically might *belong* to the United States, as we will see in the next chapter. Yet American managers refuse to acknowledge these endangered birds or to consider them in any formal sense while creating and implementing management plans. In fact, the United States could shut off the water to this marsh and these birds could be cut off at any moment. These endangered birds have no official clout to prevent that.

In other words, these rails are living creatures around which several environmental issues intersect and take shape. They are an endangered species that deserves protection and restoration—raising the related question of how best to protect and restore the delta of the Colorado River. The key to their abundance here in the delta of Mexico lies in the most precious commodity in the arid West: water. And this is a transboundary issue, since the water that sustains these birds comes from the United States. The rails illuminate the central themes of this book: the need to conserve the neglected delta of the Colorado River, and the concurrent need for a new, binational environmental vision whose unifying theme is water.

The delta has always been a unique region of North America. It is now supposed to be one of the most abused landscapes on the continent. Yet, like our burgeoning knowledge of the Yuma clapper rail populations, the dimensions of life in the delta are slowly coming into view for us.

Katán and I are sitting in a canoe smack in the epicenter of this population. The place we have discovered would turn out to be one of the best places in the delta to see Yuma clapper rails.

As Katán and I watch the Yuma clapper rail in the small clearing amidst dense tules, we are stunned by the moment. To give you an idea of how rare this moment was, you should know that no one has yet found a nest of these rails in the delta.

And there was something else.

If you attend to nature, give yourself over to it, put yourself in some sense on its level and search it carefully, you will discover these moments in which the world and its creatures reveal themselves to you. Often I am astonished by how little we really know about the world, how opaque it seems to most of us on a daily basis. It's a world of secrets and concealed truths. Like the hidden rails everywhere calling around us, it is a world of invisible presences. The German philosopher Martin Heidegger speaks of moments in which the world seems to open up to us and disclose itself. He describes them as "clearings" and "uncon-cealments" in which we can glimpse a new mode of "being in the world." He writes, "Wherever man opens his eyes and ears, unlocks his heart, and gives him-self over to meditating and striving, . . . entreating and thanking, he finds himself everywhere already brought into the unconcealed" (Bate 2000, 268).

Out of such experiences, he says, we learn a new mode of being in the world, which he calls "dwelling."

There is another delta, a new delta, emerging. It offers the possibility of a new ethic in our relations to endangered animals in the Southwest and in our relations to water in the Colorado River. It offers as well the possibility of a new politics, one that is more inclusive in its sphere of care and concern for both people and animals.

As Katán and I watch the rail next to our canoe, the bird walks behind me toward Katán. I turn slowly to watch it and glimpse Katán's face, beaming with joy. It's an extraordinary moment, which Katán and I share with the rail.

Then I hear Katán whisper something.

"Va a subir." Katán's almost laughing at the absurdity of it. "It's going to climb into the canoe!"

It doesn't, of course. But it is that close to us, right beside the canoe. I am taking pictures, but by this point, the rail is so close I can't even focus on it.

The endangered rail turns around and heads back toward the reeds. It stands for a moment at the edges of the stalks of the cattails, back to us, butt feathers flicking. The bird vanishes into the marsh. Katán and I erupt in a round of laugh-ter and joy and satisfied smiles. Then we do exactly what people always do when they have a close encounter with a remarkable creature. We talk about it. We share it. We tell each other the story of the rail that almost climbed into the canoe, several times.

This rail is the lens through which I first begin to see more deeply into the delta. As is so often the case for me, an animal has drawn me into a place, helped me to see and value the land more intimately.

How it is that this rail could come to be here, in this exact place, in a trashed-out and supposedly devastated delta? The answer to that lies in the drainage ditch we have just canoed and in the miraculous marsh that is only one hundred yards away, into which we canoe that happy evening just as the sun sets and the sky flames into a gorgeous desert orange.

TWO

# Green Lagoons

THREATENED AND RARE SPECIES: Black rail
PLACE: Ciénega de Santa Clara

IN THE AMAZING stillness and near-perfect calm of the morning, a full moon hangs low above the cattails, slowly falling. Below, the image of the moon floats serenely on the mirrored water of the marsh.

We are already out in our canoe in the morning's most magical hour, between the first hint of light and the actual rising of the sun above the horizon. Not a breeze disturbs the peacefulness that lies upon the marsh. We glide across the water, and I can suddenly understand why another desert culture, the Hebrews, pictured God brooding upon the "face of the waters" in the original act of creation, the first dawn. Still waters. I take them in. I let myself—my feelings and my imagination—expand into the huge marsh and reflect back to me like the moon on the water.

In the stern of the canoe, Juan Butrón—Katán's dad—steers us through the marshy maze of large lagoons and narrow channels. The morning in early June is warm already, since the temperature overnight did not get below about 85 degrees Fahrenheit. Juan knows the routes intimately, probably better than anyone.

We are looking for a bird that is even rarer than the Yuma clapper rail. When we conducted the clapper rail survey, we also played tapes of a tiny little rail called the black rail (*Laterallus jamaicensis*). This rail is much more difficult to find, and would offer even greater testimony to the value of this Mexican marsh. We heard two responses to this tape very near the mouth of the Ciénega's Welton–Mohawk Canal—very close, in other words, to the opening in the marsh where Katán and I discovered the Yuma clapper rail. Today Juan and I are

◄ *A full moon hangs over the reeds of the Ciénega, reflected on the still waters of a green lagoon.*

▲ *This immature green heron was so focused on its fishing, it ignored us as we watched from one of the clearings in the reeds.*

*Juan Butrón knows the Ciénega perhaps better than anyone. He has lived here since before the marsh was born. He calls it part of his family.*

canoeing to this same place in search of a black rail.

Juan grew up in the delta of the Colorado River. He calls himself *puro cachinilla*. *Cachinilla* is the local word for arrow-weed, a bushy tree also native to the desert delta and out of which early settlers built stick-and-mud houses and stick fences. Locals born in the delta call themselves *cachinilla*, signifying a native connection to the delta. It also suggests that they are tough as the tree itself, able to survive and thrive in this fierce climate and dry desert.

Juan is about fifty years old. Almost thirty years ago, he moved out from a local town in the lower delta and onto the nearby *ejido*, called Ejido Luis Encinas Johnson. At the time, the marsh did not exist. It was empty land in the desert. Juan and his wife began a family in the *ejido*. Katán was their first child, born in the mid-1970s. Soon they had five children, two boys and three girls.

About the same time, this marsh was born.

Juan calls this place his *tereno nativo* or native plot. "I brought Juanito—Katán—out on these waters when he was one year old," he tells me. The lives of his family and the life of this marsh are intertwined. For Juan, the Ciénega is part of his family's history. More than that, it is part of the family itself. Paddling over the smooth waters, he tells me that the Ciénega is to him like one of his children. "Como un niño."

Juan's feeling for the marsh is profound, paternal, and fierce. He is a handsome man, thin and distinguished looking in an old-fashioned way. His graying hair is thick and wavy, like a matinee idol in the silent era. He has spent his entire adult life exploring and defending the Ciénega de Santa Clara. No one embodies these wetlands in the delta more fully than Juan Butrón.

Through Juan these wetlands have also become one of my favorite wild places in North America. And they are almost completely unknown outside of the local region. In northern Mexico, a *ciénega* translates literally as "hundred waters." It signifies a wetland, a marsh. The Ciénega de Santa Clara is the biggest wetland in the entire state of Sonora, Mexico. It is also one of the biggest and most important wetlands on the entire Colorado River watershed—from Wyoming to the Gulf of California. Those who have gotten to know this magical place over the last few decades argue that it is one of the most spectacular and important wetlands in all of North America.

The Ciénega de Santa Clara is a vast island of water in the huge sea of Sonoran sand. These "hundred waters" cover about fifty thousand acres. The upper twelve thousand acres or so are marshy and covered in thick vegetation, a dense mat of cattails interspersed with bulrushes. The water is shallow, never deeper than about a meter, a maze of green lagoons in the midst of a brown desert.

An oasis in the desert, the Ciénega is a magnet for birds and wildlife, a crucial refuge for them. Osvel Hinojosa's survey of Yuma clapper rails demonstrated that by far the largest number of this endangered species occurs here. Studies have only just begun to document the immense natural richness of the Ciénega. It is home to an endangered fish, the desert pupfish, and is used by a still-undetermined number of other endangered birds such as brown pelicans. For bird-watchers, it's a paradise of more than two hundred species of birds. (The list is still being compiled.) Up to 300,000 waterfowl winter in the Ciénega, and shorebirds depend upon it for both breeding and migration.

In 1993, the Mexican government recognized the value of the marsh by making it a major part of the Reserva de la Biósfera Alto Golfo de California y Delta del Río Colorado. The marsh has also been listed under the United Nations Ramsar convention, which designates wetlands of major international importance. José Campoy, the director of the biosphere reserve, says the Ciénega is one of the most important wetlands along the entire Colorado River.

Aside from its astonishing natural values, what makes the Ciénega even more magical is that in a very real sense, it should not even exist. It is quite literally an accidental refuge, and the history of this wetland is one of the strangest conservation stories in North America. The Ciénega's creation is an amazing tale of historical mistakes, bureaucratic miscalculations, and international water politics played at the highest levels and for very high stakes. The marsh may well be a jewel in the desert, but it occupies an even more precarious position—exposed and vulnerable—in the always-bizarre world of western water wars. Neither God nor nature created this Mexican marsh. Ironically, that other god of the West created it, the United States Bureau of Reclamation.

How strange is this marsh? It exists in Mexico, but the water that fills it might belong to the United States. What also makes the marsh so magical is that it's one of the best-kept secrets on the Colorado River—despite the fact that you could create a second river out of all the ink that has been used to write about every aspect, every nook and cranny, of this once-great river. No paved roads lead to the hidden Ciénega. It is not visible from any highway. Local Mexicans know about the marsh. But even though it's only about thirty-five miles below the U.S.–Mexican border, Americans are mostly oblivious to it. On many days when Juan and I have been on the marsh, we have had it all to ourselves—fifty thousand acres of water and birds and solitude.

It's not surprising that so few Americans know about the Ciénega. The delta has always occupied a blind spot in the collective consciousness of Americans. It is a landscape of neglect and oblivion. The delta was completely ignored as America went on what author Marc Reisner calls, in his polemical and indispensable book *Cadillac Desert: The American West and Its Disappearing Water,* "a half-century rampage of dam building" (Reisner 1993, 51). Little or no thought was given at the time to what might happen to the delta in Mexico. Since the first dam went in at Boulder Canyon—Hoover Dam, 1935—about eighty dams and diversions have been built on the rivers of the Colorado River watershed. In the process, the flows of the Colorado River to the delta and to the Gulf of California were completely cut off. The last great dam went in at Glen Canyon. For almost twenty years, from 1963 to 1981, its reservoir—Lake Powell—slowly filled. All the water in the river was captured and the delta dried up.

In twenty-four of the last forty years until 1999, only 2 percent of the water in the Colorado River reached the sea. The 120-mile stretch of the river through the delta in Mexico was hung out to dry. The first dams were built, of course, before environmental impact statements were required. Despite the unregulated, unexamined growth of that time, we still need to ask an enormous question: How could the environmental devastation on the delta have been so completely and, it appears, so willfully ignored as the dams were built?

During those energetic years of damming, American attention was focused almost exclusively on building a civilization in the desert out of the water taken

from the river—an historical imperative that would tolerate no second thoughts. The Colorado River now supports about 30 million people along its 1,400-mile length, and the number is growing. The Colorado River irrigates about 2.5 million acres of farmland in the U.S. portion of the delta. In the Mexican portion of the delta, another 500,000 acres have been converted to farmland. The entire flow of the Colorado River has been appropriated for people under a principle of "total use," in the words of Donald Worster in his magisterial book, *Rivers of Empire: Water, Aridity, and Growth in the American West.*

It's a thirsty civilization living in the hottest and driest part of North America, and that civilization has turned the Colorado River into the world's greatest plumbing system. Turn on a tap anywhere from Arizona to California and you will see the new river rushing from your faucet, water you depend upon for your life. By one estimate, every drop of water that actually reaches the lower part of the river has already been used three times. One of the central facts about the modern Colorado River is that it's heavily overallocated for human use. We already support more people in the arid West than we have water for.

Conveniently ignored, the delta paid the price for this frenzy of growth to the north. Once the delta covered 1,930,000 acres. Once it was one of the most remarkable of natural phenomena—a lush and huge oasis in the midst of one of the driest deserts in the world. Because it was such an intimidating desert, few people braved it. One of the few who did was Aldo Leopold, the great conservationist. In its pristine state, it was one of the most spectacular desert

*Dramatic light reveals the labyrinth of water and reeds in the Ciénega.*

deltas in the world, rivaled perhaps only by the great Nile delta, to which it was always compared.

After the dams, what remains is a remnant delta, truly one of the saddest and most endangered complexes of ecosystems in North America. The "wild" delta now is confined to about 150,000 acres, entirely in Mexico, beyond the agricultural flood-control levees that line the southern agricultural boundaries of the delta. The wild delta has shrunk to about only 8 percent of its former expanse.

As the dams went up and their reservoirs slowly filled, the delta came to be seen as a biotic death zone, withering for lack of water. What seems also to have happened is that the delta was not only ignored; as it dried up, it came to be written off by those who should perhaps have known better. The delta soon became a symbol of the environmental devastation wreaked by the dams on the river. That very symbol paradoxically contributed to American lack of concern for the delta and for the lives of those who do live there.

Even as the Ciénega developed in the delta, writers continued to describe the place solely as degraded and devastated, a wasteland. This was a new kind of blindness. Falling into a trap of their own preconceptions, these writers constructed images of the delta that still make what is really going on down there invisible to us.

Consider the way that even environmentally concerned writers have treated the delta. The most definitive popular book on the Colorado River, Philip Fradkin's *A River No More*, established the trend in viewing the delta as dead. First published in 1968, at the low point of river flows into the desert, the book was reissued in 1995. By then Fradkin might have seen more clearly. Yet he continued to represent the delta as all but buried. He calls his final chapter "Ends: Death in the Desert." He describes the delta as a barren, "illusionary wasteland." The delta, he says, "is symbolic of what the West would become" (Fradkin 1996, 322).

One of the great books on water in the West, and perhaps the most quoted, is Marc Reisner's *Cadillac Desert*, which takes the same approach. Even in the revised edition of 1993, only two references to the Mexican delta of the Colorado River are found. While Reisner's focus is on water in the United States' West, the exclusion of transboundary water issues is symptomatic of a general national disregard for the end of the Colorado River. In his treatment of the delta, Reisner effectively signs its death certificate: "Amid the salt-encrusted sands of the river's dried-up delta," he writes, the West begins "to founder on the Era of Limits" (Reisner 1993, 121). Summing up the ecological state of the delta, he dismisses the entire region. "The Colorado delta is dead," he states flatly (485).

Fradkin and Reisner are two very good writers. They are part of a long tradition in which the delta has become the prime symbol of the ecological disaster of the West. A National Geographic book (and article) from 1992 perpetuates the myth. The classic photographic image is of a trickle of water drying up in the sand. Less than five miles south of the international border, reads the photo caption, "Bird tracks and a slowly sinking pool of water in Mexico mark the end of the once mighty Colorado River" (Carrier 1992, 161).

I could name many other books, many other writers. The problem here is that these images create a mental trap: If the river is dry and the delta is dead, it is also past hope. The battle has been waged and lost. We might as well forget about the delta and its future. The delta is an environmental casualty of our water-guzzling culture. While this rhetoric is useful as an indictment of the program of American dam building, the bitter irony is that this symbol of the dead delta sacrifices the delta once more.

Of course, the stereotyped images of the delta are not entirely wrong—clichés and stereotypes never are. The delta *is* a profoundly damaged landscape. But as long as we pronounce it dead, the delta as wasteland becomes a self-fulfilling prophecy. There is no reason to change our habits or behaviors. We can continue to exploit the delta's water without guilt.

More important, though, the stereotype is inaccurate. Even as the cited books and many other books were being written and published, the Ciénega de Santa Clara was home to a wealth of rails and migratory birds. Juan Butrón was living beside the Ciénega. Neither he nor his family was dead. The delta was not dead for them, nor for 200,000 other people who live in the "wild" part of the delta. The delta was not dead for the creatures that came to depend upon it. Endangered Yuma clapper rails had already colonized the Ciénega when both Fradkin and Reisner were writing. Hundreds of thousands of migratory ducks and geese had discovered the marsh. Coyotes were already fishing for crawfish there.

The Ciénega is an example of the accidental regeneration in the Mexican delta, as well as of the delta's ecological potential. It is a microcosm of the larger delta. Understanding how the Ciénega was created takes us into the strange new world of water and the delta, which serve as a crucible of the strange world of water issues in the arid West. The Ciénega story also shows how this tender recovery is already in jeopardy.

There is no better place to understand the Ciénega than where I am headed this morning with Juan. In our travels, we will encounter abundant proof of the inspiring biodiversity of the delta. After negotiating several lagoons, we arrive at the spot in the northeastern part of the Ciénega where the Welton–Mohawk Canal empties into the marsh. It forms a delta of its own as it spews water and plastic trash—a wide fan of rich mud over which sweeps the discharged water from the canal that comes from the United States.

This exact spot is my favorite location in the entire delta. It is very close to where I had seen the Yuma clapper rails, just up from here in the canal. It is always full of birds and wildlife. Regularly, we see raccoons in the edges of the tules. This morning, we arrive to spot a coyote standing up to its belly in the

*A short-billed dowitcher, a shorebird, stretches its wings in anticipation of a long migration. Birdlife is abundant in the Ciénega, which has become a crucial link in the migration of West Coast birds.*

water. It splashes noisily away as we sneak into view. Pelicans—both brown and white—float on the water. A short-eared owl flies low over the marsh, heading to its roost. At the delta of the Welton–Mohawk, black-necked stilts and graceful, lovely American avocets are already busy feeding. A flock of short-billed dowitchers rests in the exposed mud. The canoe is silent. We are cautious. We are able to get very close, within ten feet of the shorebirds. I photograph several of the resting birds.

"*Mansitos,*" Juan whispers. Tame little things.

Behind them is a specialty of the area, a rare gull-billed tern. I look carefully among the ring-billed gulls. I have also seen yellow-billed gulls here, not to mention both sora rails and Virginia rails. Deltas are always rich spots, and this delta within the delta is the epicenter of the region's fabulous bird life. While we are watching the shorebirds, a tricolored heron flies in— the first time this bird has been recorded in the delta and a sign of how we are only just beginning to document the area's hundreds of species of birds, along with its superb biodiversity—all in the heart of what is supposed to be a "dead delta."

But we are here for a specific reason. We both climb out of the canoe, sinking up to our calves in the smelly muck. We get ready to drag the canoe over the mud and up through the tules, toward the small clearing along the canal to see what might be waiting for us there.

When Ed Glenn first saw the huge sheet of water, he must have thought he was seeing a mirage. Poking around in the delta, he had followed the long concrete canal, the Welton–Mohawk, to its discharge point in 1979. Everyone else had given the delta up for dead. Yet here he stood, looking at a shallow sheet of water covering thousands of acres.

In the revival of interest in the delta and the growing battles to save it, this remains one of the key moments. Where Ed Glenn stood is also in one of the key places in the delta—very close to where I was bird-watching with Juan Butrón some twenty years later.

A botanist at the University of Arizona's Environmental Research Laboratory in Tucson, Glenn describes himself as "one of the few people studying the Colorado River who have actually been to the delta." He is soft-spoken, almost shy. He has a youthful, chubby face, mischievous eyes, and a Puckish wit. He is, perhaps, the dean of scientists studying in the delta. Certainly he has been at it longest and is one of the delta's most effective advocates. His advocacy for the delta can be traced to that moment in the late 1970s when he found the new lake in a "dead" delta.

In 1979, Ed was in the delta looking for halophytes—plants especially adapted to salty environments. The delta of the Colorado River, in Mexico, is a particularly good place to study them. The original delta had a vast intertidal plain, some 300,000 acres where salt-tolerant plants grew. With all the agricultural fields upstream, much of the delta was being poisoned by salts as well. In fact, you can find big "puddles" of salt, inches thick, just sitting on the mud. For a specialist in salt-tolerant plants, he was exploring a salty paradise.

Ed had heard that new water was being sent down into the delta, and he decided to look for it. He drove to the concrete-lined canal south of San Luis Río Colorado and followed the drainage ditch to the end. What he found was a new lake. The water was not deep and was still clear of plants. It shimmered under the desert sun like an illusion.

He was witness to the infancy of the Ciénega de Santa Clara.

He did not discover this new lake, he is quick to make clear. Local folks were already making use of it: "There were lots of Mexicans already there, fishing, camping, having family picnics. Mullet had already entered the marsh from the sea in very high tides. Kids were even swimming in the water." It is entirely possible that Juan Butrón and his family were one of the Mexican groups enjoying this new lake when Ed Glenn first squinted into its shallow depths.

Ed knew the lake would have to change the ecology of the area. He filed the phenomenon away mentally. Then several years later, he says, he decided to go back and see what it was like. The lake had become a wetland paradise, a refuge of lagoons and green cattails for people and creatures, in an area that before had been nothing more than a parched and salty mudflat.

How this new lake and marsh came into being is one of the strangest stories in the history of the Bureau of Reclamation, an agency noted for strange stories. The genesis is the result of a huge battle in the 1960s and 1970s between the U.S. and Mexico over water allocations from the Colorado River. Long before North American settlers first diverted the Colorado River into the Imperial Valley in 1901 for agriculture, the two countries had been fighting over the river and its water. In 1944, they signed an international treaty dividing the waters—

a story we will return to in later chapters. The United States secured rights to 90 percent of the Colorado's flows: 16 million acre feet (maf). Mexico receives 1.5 maf, plus another 200,000 af in years of surplus water in the watershed (say, from heavy snows in the winter).

When you "talk water" in the West, you talk in "acre foot" units—the primary standard of measurement. Specifically an acre foot is the amount of water required to cover an acre of land to the depth of one foot. This amount is customarily said to be able to support a family of four for about one year. It's also equivalent to about 326,000 gallons. Perhaps the clearest way to think of it, if you are not a water engineer, is this: one acre foot of water will cover almost an entire American football field with one foot of water.

Nothing in the treaty with Mexico, however, stated explicitly that the water had to be usable.

This oversight came to a head in 1961. As the water of the Colorado River is pumped onto farmlands and recycled over the course of its march downriver, it accumulates salts. They leach out of the soil with each use. The salts in the water increase from about 50 parts per million (ppm) at the headwaters to 879 ppm at Imperial Dam. This growing salinization is a major problem in the management of the river under the best of circumstances, since it corrodes plumbing, damages waterworks and municipal delivery systems, decreases crop yields, and makes conservation harder. Every year the Colorado River carries 9 million tons of salt. A single part per million increase at Imperial Dam is estimated to cost the Bureau of Reclamation an extra $108,000 in operating expenses—and that cost rises every year. About 1.5 million tons of salt have to be removed from the water at present to reach municipal consumption standards. One estimate puts the cost to the Lower Basin states (California, Arizona, and Nevada) of removing salts from the Colorado at $750 million per year. Salt is both economically and environmentally very expensive.

In the 1960s the salt content of the water delivered by treaty to Mexico shot up to over 2,000 ppm. It was like pouring poison on Mexican crops, and the anguish of Mexican farmers soon exploded into a full-fledged international incident.

The spike in salt was caused by a huge U.S. irrigation project. The U.S. government had authorized the Bureau of Reclamation to go forward with a project in the Welton–Mohawk district along the Gila River, not far from Yuma. Below the surface of this land was a perched aquifer sitting atop clayey soil. The aquifer contained heavy salts. When farmers began to irrigate the lands, the runoff water picked up these salts from the aquifer. The water that was dumped back into the Colorado spiked to over 6,000 parts per million (ppm), causing the water in the river to hit saline levels of 2,000 ppm.

When farm production crashed in Mexico because of this sharp increase in salinity, the farmers took to the streets in protest. By the early 1970s, Colorado

River water salinity had become a national issue in the Mexican presidential campaign. Mexicans demanded that the 1944 water treaty be revised to require that the United States deliver fresh water, not salt.

After negotiations at the very highest levels between the two governments, a modification to the treaty was approved. These "modifications" are tantamount to amendments; in the legal language of the treaty they are called "minutes" and they reflect decisions made during meetings of the International Boundary Water Commission. The agreement on the salt content of the water delivered to Mexico is called Minute 242. For Mexicans, it is one of the most important minutes yet to the original treaty, and it established salinity guidelines for the United States' deliveries. Essentially, the United States must deliver water that is no more than 115 ppm more saline than the water used by American farmers near the Imperial Dam—about 1100 ppm.

However, promising clean water is one thing. Delivering it is another. To meet the new treaty requirements, the United States had to get rid of the salty water coming out of Welton–Mohawk Irrigation District. But that salty water could not go back in the river. How could the United States accomplish this?

The Bureau of Reclamation followed a two-part strategy. First, they built concrete-lined the 50-mile Welton–Mohawk canal south from Arizona into Mexico. It is officially called the Mode Outlet Drain Extension Bypass, or M.O.D.E. The salty wastewater from the Welton–Mohawk Irrigation and Drainage District flows through this canal and is discharged into the lake at the end of the canal.

Second, Congress authorized the construction of a huge desalting plant outside of Yuma. Using a reverse-osmosis process that filters the water through ten thousand pipes, the Yuma Desalting Plant would clean up the water on the outskirts of Yuma. The plant would then dump this desalinated water into the Colorado River, and it could once again be part of the deliveries to Mexico. The waste brine from the process would go back down the canal to Mexico. That was the theory and the plan. The Bureau of Reclamation spent about $350 million on the canal and the Yuma Desalting Plant. The elaborate scheme never came to fruition.

Why? The canal was complete in 1977. Soon thereafter, 120,000 acre-feet per year of brackish water were being pumped down into the Mexican mudflats of the delta—the equivalent of about 80,000 Olympic-sized swimming pools. The current salinity is about 3,500 ppm—far too salty for agriculture, but merely brackish for swampy plants that thrive in estuarine backwaters. For cattails, the salinity is perfect.

As the water flowed onto the mudflats, it first created a huge shallow lake—which Ed Glenn saw about a year later. Then the cattails took over. This was not part of the plan. In fact, I am not even convinced that the canal that the bureau built was supposed to stop where it does, in the middle of the delta. The draft Environmental Impact Statement (EIS) that the bureau prepared says the canal is to go all the way to "Santa Clara Slough," an arm of the sea reaching up into the mudflats. That would mean the waters would have been effectively dumped into the Gulf of California. But Ed Glenn says that the description of the place in the bureau's plans indicates that they never intended the canal to go all the way to the slough. They did expect, though, that high tides would wash out this salty water and carry it to the sea.

In any event, the dumping of this water into the delta was supposed to be temporary. But then the Yuma Desalting Plant went through interminable delays. It was not finished until 1992. In addition to the cost of building the plant and canal, estimates put annual expenses of running the plant at $24 million. Except for one brief period, the plant has never been used. It is now in mothballs while the bureau and the Welton–Mohawk Irrigation and Drainage District try to figure out what to do with it.

Meanwhile, the water continues to flow to the delta, sustaining the Ciénega de Santa Clara that it created. And the Ciénega has no official habitat protection whatsoever in the United States. The Mexican government included the Ciénega when it established the biosphere reserve in the delta. But the legal force of this reserve status in the United States is not certain. The United States

*The strangest of natural phenomena, the Ciénega is a wetland in the desert. These wetlands in Mexico were created by accident by the U.S. Bureau of Reclamation. Aldo Leopold explored this exact spot by canoe almost eighty years ago, making it famous in his essay called "The Green Lagoons."*

could decide at any time to cut off the water, bring up the desalting plant, or use the water for some other purpose (it could, theoretically, use the water for power generation, or cooling a nuclear power plant). As the Director of the Lower Colorado River Bureau of Reclamation, Robert Johnson, reminded me in an interview, the water in the Ciénega de Santa Clara "actually belongs to the United States." The $360 million Yuma Desalting Plant looms over the Ciénega like a very big threat. If the desalting plant were to come on line—and there are studies under way right now to decide what to do with it—the flows to the Ciénega would be drastically reduced. Worse than that, Ed Glenn conducted studies showing that the resulting heavy brine discharge would virtually destroy the marsh ecosystem.

For now, the M.O.D.E. Bypass stands as perhaps the Bureau of Reclamation's most beautiful mistake ever. The bureau certainly did not intend to create a wetland so important and spectacular that it could become an internationally acclaimed wild area and biosphere reserve. That it intended to slough heavy brine in the delta shows that the bureau conceived of Mexico's delta as a kind of dumping ground—a place where it could flush the toilet and send our waste.

According to Ed Glenn, the marsh began to prosper almost as soon as the water arrived in the late 1970s. Satellite images show what happened, as do the Bureau of Reclamation's own studies.

"The cattails started immediately," Ed says. "It wasn't slow. There was a big area to go from mudflat to 15,000 acres of vegetation."

In 1972, while Lake Powell was filling upriver, satellite images show that the region that is now the Ciénega was dry dust. Just five years before the Welton–Mohawk Canal was finished, and shortly before Juan Butrón moved into his *ejido*, only 1,000 acres of wetlands existed in this area. By 1975, the delta was near its low point. Only 75 acres of cattails survived in the area, fed by brackish water seeps and agricultural runoff. But these would be enough to seed the regeneration. The Welton–Mohawk Canal started discharging in 1977.

Only eleven years later, in 1988, about 7,000 acres of cattails were crowding the shallow lake. In July of 1993, according to Bureau of Reclamation scientists, the total had reached almost 11,000 acres. For the last decade, that amount of vegetation in the upper part of the Ciénega has remained more or less constant.

These 50,000 acres of reeds and water have become a crucial and beautiful part of the wild Southwest. One estimate suggests that as many as 400 plants and animals depend upon the Ciénega. Perhaps more than anywhere else in the delta, the Ciénega has become a symbol of hope and a rallying point for preserving the delta.

José Campoy, the biosphere reserve director, puts it this way: "The mission here is to let people know that a protected place for the cycles of life exists. Let's recognize that it's not just a desert down here. With the help of the United States, the challenge is keeping it protected."

The huge lesson is that the restoration was easy. No one had to do anything. Just add the water. Nature did the rest. It is a lesson repeated in several places throughout the delta. It defines the conservation imperative for the delta: find water.

The problem is that the United States covets this water. Robert Johnson told me that the waters delivered to the Ciénega are currently "significant to the United States as a potential way for meeting treaty obligations, and water users in the United States feel they have a right to these waters." The bureau is currently looking at the possibility of future operation of the Yuma Desalting Plant. It is also in the process of transferring the title of the facilities in the Welton–Mohawk Irrigation and Drainage District to the district itself. The implications of this transfer on the Ciénega are not clear, and they hang menacingly over the marsh's survival.

As Robert Johnson ominously puts it, the Ciénega in Mexico is "not a recognized consumptive use of Colorado River water."

Still, even the bureau has come to realize the environmental significance of these waters and the agency is trying to work out informal and "creative" solutions to the problem of guaranteeing water to the Ciénega—though none has yet emerged. The proposed solutions have mostly been technical, such as pulling water out of the groundwater bank for the Ciénega, or using Yuma valley drainage water for the Ciénega, or even paying users to forebear use.

As I write, the Ciénega's future is by no means assured.

Sunk in the muck of the marsh, Juan and I plunge up the Welton–Mohawk Canal, headed for the same spot that Katán and I discovered. We are curious as to what might be there, and continue to hope for a sighting of the black rail. We have to get out of the canoe and pull it through the canal. Now we are up to our thighs in mud and water, but it is the only way to get the canoe through the tight opening in the tules. Cobwebs and puffy brown cattail seeds soon have us covered with something like fur.

Juan jokes that we look like *tigres*, jaguars.

When I ask him, Juan tells a different story about the creation of the Ciénega from the one told by scientists and the Bureau of Reclamation. For Juan, the Ciénega is not a story about a place where water is dumped and then retrieved, or where data is collected for engineering projects. It is the story of his home and family—not so much local knowledge as lived knowledge. It is the story of a place that he loves and that has special meaning for him, his family, and his *ejido*.

Before the marsh was created, Juan says, he was able to walk the eleven kilometers from his house to here, the mouth of the Welton–Mohawk Canal. To get to the Ciénega from his house in the Ejido Luis Encinas Johnson, you have to drive about eight kilometers. Leaving the *ejido* on a dirt road, you cross several planted fields and a dense stand of *cachinilla*, arrow-weed. Gambel's quail scatter in front of the car. Then it's over the levee that protects the farmlands and separates them from what's left of the "wild" delta. For several miles you drive through an empty flood plain, cracked in places like dry skin, salted out and treeless. If the tides are very high, or if an unusual winter rainstorm hits, the road turns to a quagmire.

There is no gradual transition to the Ciénega. One minute you are charging through a rutted track in the dirt. The next you have arrived at the marsh with a small dock for launching canoes and *pangas*, or small boats. This contrast between the dreariness of the desert plain and the sudden lushness of the marsh could not be more striking.

I ask Juan if anyone—either from the United States or from Mexico—consulted him or his *ejido* about dumping wastewater here in the delta.

"No," he says, and laughs. He says they knew the Americans were building something in the delta, a long canal. But they had no idea what it was being built for.

"We thought it would go all the way to the Gulf. We didn't know the water was coming to us at all. Just all of a sudden, they gave us a bath."

"*Nos bañó,*" he says in graphic Spanish.

When he first moved to the *ejido* a few years before, there was no water here. It was damp right where we are standing, he says.

"There were a few cattails," he says, "plus some salicornia. Mostly, it was mud."

He says that they used to farm the land around the current dock that reaches into the Ciénega. They planted *cevala*—barley—and sold it to the Tecaté *cervecería* near Mexicali. For the *ejido*, *cevala* was an important cash crop, though not much any more.

About two hundred people live in the *Ejido* Luis Encinas Johnson. Among the families they have divided about 6,029 hectares, or about 15,000 acres, which they still farm in cotton, onions, and other cash crops. But it's a tough living on the *ejido*. Juan himself works 16 acres and made $1,400 in seven months the year before.

Like Katán, many people work in the *maquiladoras*. That's hard work, if you can get it, and not much pay. Buses come through the *ejido* to pick up the workers early in the morning—3:30 A.M. for the Sony plant in Mexicali, 5:30 for the Daewoo plant in San Luis Río Colorado. When they can, people supplement those earnings with what they can earn fishing in the Gulf and doing odd jobs for the biosphere reserve.

As the water began to flood their *ejido* lands, Juan says, the residents did not get indemnified one peso. Technically they still own land under the Ciénega. When the Mexican government incorporated the marsh into part of the biosphere reserve and named it the Ciénega de Santa Clara in 1993, again the people were not compensated. Scientists, he says, did not really begin to come here until the proposal to protect this habitat began in earnest.

Juan looks around at the marsh. "Son terenos de nosotros," he says emphatically. He means it literally. This is our land. It just happens to lie under a meter of water.

Small wonder that Juan and his *ejido* have a proprietary concern for the Ciénega and its health. They have come to know the marsh intimately, its mazes and its creatures. Soon, with the help of the Sonoran Institute in Tucson and Pronatura Sonora in Mexico, they have recently begun to develop an ecotourism business based on tours of the marsh. So far, this destination is anything but luxurious. It has been a struggle to get the infrastructure in place that would give Americans, the target market, the comfortable accommodations they demand. Capital is the problem, capital to build lodging and provide food. Ecotourism offers them the greatest prospect of raising their standard of living. For Juan and the others, the Ciénega's future is their future. The growing environmental consciousness in the *ejido* is a sign of a new awareness in the delta as well.

As Juan and I talk, I try to imagine this area as a mudflat. It's hard to do. Reeds and birds surround us. Yellow-headed blackbirds are croaking from their perches on the tips of the cattails. Marsh wrens, tiny birds whose short tails are sweetly cocked in an attitude of complete earnestness, strain to croak their unmusical songs. In the mud, sanderlings and western sandpipers scurry and peck for food like wind-up toys.

As we approach the clearing in the tules, we stop talking. We don't want to scare whatever might be out in the open.

There is no black rail. But on the edge of the mud, a juvenile green-backed heron with maroon streaks on its chest is frozen in an attitude of complete concentration. Herons are among the most admirable of birds—I envy their focused attentiveness. This heron's entire life seems distilled in its unwavering stare. It jabs a minnow so quickly I almost can't see the strike.

We emulate the heron in attentive concentration. The clearing becomes a place of animal revelations. Juan imitates the sound of a clapper rail. Nothing. But a sora rail comes out, feeding. Then a second sora. About the size of a fat dove, they are a midnight blue and brown, with dark lateral streaks on their butts. Both are in breeding plumage, with brilliantly yellow beaks that seem almost garish on such heavily camouflaged birds. One of the soras walks into the clearing, working the muck and algae for food. It approaches almost to my legs, which have suctioned slowly into the muck past my ankles.

I look back at Juan, smiling, and he nods to the right, immediately beside him. A Yuma clapper rail steps through the tule stalks and then out into full view. Then another. We have four rails in full view, walking right up to us.

The Ciénega is a
haven for birds in
the rail family.
In addition to this
feeding sora, the
Ciénega also is home
to Virginia rails and
the extremely rare
California black rail.

After nearly two hours, we get back in the canoe and slide down the canal. Almost ready to emerge again into the open lagoon of the marsh, Juan whispers my name. I turn around. He tells me that a small rail is just inside the cattails. We grab cattail stalks to hold the canoe in place in the current, straining to get a good look at the bird. We are hoping that it's a black rail. I spot the bird as it makes a short flight behind Juan across the canal. It's a Virginia rail, another small rail with a long orange beak and cinnamon chest.

Back in the open Ciénega, we are ebullient from all the amazing birds we have found skulking in the marsh. The black rail has eluded us. So we agree as we paddle that we will look for them tonight in another part of the Ciénega—a place where black rails are thought to have been seen.

As we paddle back to the dock, through the labyrinth of canals and lagoons, the slow journey allows time to reflect. I try to envision the delta before dams. From the few good reports we have from that time, the Ciénega is a superb example of the delta's original richness. There is no better witness to that richness than the great environmentalist, Aldo Leopold.

Aldo Leopold put his canoe into the waters of the delta at almost exactly the same spot where the canal now empties into the marsh. The location is layered with natural and historical value. That Leopold chose this spot makes the Ciénega one of the most meaningful places in the environmental history of North America.

On October 25, 1922, Aldo Leopold crossed the United States–Mexico border south of Yuma, stopped for a beer in San Luis Río Colorado, and headed south into the delta. With him were his brother, Carl, and his dog, Flick.

With a canoe strapped to their Model T Ford, they were embarked on a "shining adventure." Aldo Leopold drove into Mexico's Colorado River delta on what he called a "voyage of discovery." He went in search of a still-wild place, a blank spot on the map. By the time he returned from their three-week excursion, he had crossed from geography into myth. Leopold is one of only a handful of early writers who offer tributes to the original wild delta. In 1922, the delta had already begun to be colonized and transformed. Hoover Dam, though, was still thirteen years in the future, the "watershed" event that would change everything on the river.

Aldo Leopold is often called the father of the American environmental movement. His great book, *A Sand County Almanac, and Sketches Here and There*, published posthumously in 1949, articulated a new ethical relationship between human beings and the land. The "land ethic," as Leopold called it, transformed humans from conquerors of the land to citizens of a much more democratically imagined land community. It is a famous book, one of the most important pieces of American nature writing that we have. Perhaps one of the ironies of American environmental history is that one of the crucial landscapes that helped Leopold discover and articulate this ethic was not in the United States, but in Mexico and the delta of the Colorado River.

His short sketch called "The Green Lagoons" describes his trip to the delta. This essay and the original journal entries he wrote, separately published in a volume called *Round River*, leave us with a sensitive record of the legendary beauty of the pre-dam delta.

The Leopolds drove south with their dog and their canoe to a small bulge in the road called Rillito, now Riito—"the little river." At this spot, in a "pretty little slough" just below Riito, they put in their canoe. For all practical purposes, Aldo Leopold explored the exact same area where the Ciénega now exists. He described the waters as having a "deep and emerald hue."

Leopold encountered a "verdant wall" of mesquite and willow and *cachinilla* all around the sloughs. The river flowed through the desert and created a paradise for any person who loved wildlife and wild places:

> A verdant wall of mesquite and willow separated the channel from the thorny desert beyond. At each bend we saw egrets standing in pools ahead, each white statue matched by its white reflection. Fleets of cormorants drove their black prows in quest of skittering mullets; avocets, willets, and yellow-legs dozed one-legged on the bards; mallards, widgeons, and teal sprang skyward in alarm. As the birds took the air, they accumulated in a small cloud ahead, there to settle, or to break back to our rear. When a troop of egrets settled on a far green willow, they looked like a premature snowstorm. (Leopold 1949, 142)

Leopold might just as accurately be describing the modern abundance of bird life in the Ciénega.

Stalking this "wealth of fowl and fish" were many predators. Bobcats were so common—he reports seeing seven—that they called one campsite "Cat Camp." Coyotes were also everywhere. They watched coyotes, he writes, "from inland knolls, waiting to resume their breakfast of mesquite beans" (Leopold 1949, 142). Around the eastern margins of the Ciénega, where *tornillo* mesquite trees are still found, coyotes still leave scat full of *tornillo* beans.

For Leopold, the great symbol of the delta as a wilderness was the jaguar. He writes that at every shallow ford they found tracks of "burro deer," or mule deer. They followed the tracks, he says, hoping to find "signs of the despot of the Delta, the great jaguar, *el tigre*." He never saw one, but claimed its "personality pervaded the wilderness."

The black-necked stilt
is one of the most
elegant of shorebirds
of North America.
They are common
nesters in the delta.

In the weeks in the slough, Leopold says his mood changed to match the lazy river in its slow slide across the flat delta toward the Gulf. He writes:

> . . . the river was nowhere and everywhere, for he could not decide which of a hundred green lagoons offered the most speedy and pleasant path to the Gulf. So he traveled them all, and so did we. He divided and rejoined, twisted and turned, he meandered in awesome jungles, he all but ran in circles, he dallied in lovely groves, he got lost and was glad of it, and so were we. (Leopold 1949, 142)

The current brackish waters of the Ciénega don't slide like the river toward the delta. The Ciénega is a marsh of cattails and rushes, not a slough in a slow desert river. The current marsh is also much smaller than the green world of the original delta wetlands. Nevertheless, the Ciénega is a modern version of Aldo Leopold's "green lagoons"—a labyrinthine wonderland of wildlife and slow time.

Until the 1930s, the Colorado River created a diverse set of ecosystems in its delta. After dropping fast and hard out of the Rocky Mountains, the great river leveled off after coming out of the Grand Canyon and began a more measured and dignified march toward the sea. By the time it reached the delta, the Colorado passed over thick sediment deposits that it had itself laid down, an enormous meander of marshes and forests in the middle of a desert. The delta itself was created by the river, built up by the river's own burden of silt, as we will see in more detail in chapter 3.

Passing through its delta—which was the size of the state of Connecticut, or nearly twice the size of Rhode Island—the river made a green belt to the sea. One reads various estimates of the original size of the green belt of the river, but since no careful records were compiled, it's hard to say for sure. But the extension of bottomland marshes and forest would have been about 10 miles wide, extending south from the present city of Yuma, Arizona, for 75 miles. That's 750 square miles of steamy green fertility, surrounded by baking desert sands. The area is about twice the size of, say, Mount Rainier National Park, or nearly half the size of the immense Grand Canyon National Park.

The huge tides of the Upper Gulf extended their saline influence about 35 miles up the river, and from the end of the green lagoons a vast intertidal plain took over. The North American continent had no other desert delta like that of the pristine Colorado River. It was utterly unique—vast, daunting, hot, and home to enough birds and wildlife to stagger the imagination. On a global level, only the Nile and the Indus Rivers were comparable for spectacle and scale.

Leopold was smart enough to explore the delta before it was lost. Very few other accounts of the pre-dam delta have come down to us, and none other as beautiful. The delta was studied for its geography, for its geology, and for its hydrology—matters of navigation, farming, and engineering. But not for its natural history. However, two other writers give us glimpses of the pre-dam abundance in the delta that corroborate Leopold's description.

One is a trapper, James O. Pattie, who journeyed down the Colorado River in 1828. He was fleeing Native Americans around Yuma. As he descended the river, he encountered an exotic abundance that beggared his powers to describe:

> There are but few wild animals that belong to the country farther up, but some deer, panthers, foxes, and wild-cats. Of birds there are great numbers, and many varieties, most of which I have never before seen. We killed some wild geese and pelicans, and likewise an animal not unlike the African leopard, which came into our camp, while we were at work upon the canoe. It was the first we had ever seen. (Pattie [1833] 1966, 143)

The animal that Pattie likens to "an African leopard" was almost certainly a jaguar, Aldo Leopold's "despot of the Delta," now long gone from the region.

About a century later, contemporaneous with Leopold, another hunter/explorer recorded a memorable description the delta. Lewis R. Freeman left an account of his experiences of the delta in *Down the Grand Canyon*. The book was published in 1930, but the trip actually took place in 1922, only shortly before Leopold's. In fact, Leopold asked after Freeman as he entered into Mexico.

For Freeman, the delta was both a "Golden Land of Enchantment" and one of the "queerest, weirdest regions that fancy can picture." Freeman's telling reports much more encroachment of people and cows and farms than does Leopold's account, which instead lays its stress on a mythic natural abundance. Nevertheless, Freeman describes the "swarming wild life" in the delta, which was a "paradise for the hunter." The river offered "an easy fresh-water route through the one-time *terra incognita* where ranged such hordes of wild life as few but the earliest of our pioneers have ever seen" (Freeman 1930, 67–8). By 1930, Freeman writes, that world is already gone.

By the time Aldo Leopold came to write his essay, "The Green Lagoons," in the early 1940s, the delta was even farther gone. Hoover Dam in 1935 was the definitive event that changed the river in the last century, announcing an historical imperative in dam and empire building in the West that no one seemed able to oppose. It sealed the river's fate—and with it, the delta. Leopold was so saddened to think about the changes that must have happened in the delta because of the dam, he said he could not bear to return.

Rather he wrote a nostalgic tribute to the lost delta, an elegy. The trip was a transformative experience for him, Leopold claimed. One of the powerful features of the essay is not the testimony to the land's abundance, important as that is, but rather the moral and psychological dimension of the landscape. The delta instructed Leopold in a new way of relating to the landscape. He describes the feeling as "sensitivity to the mood of the land":

> We could not, or at least did not, eat what the quail and the deer did, but we shared their evident delight in this milk-and-honey wilderness. Their festival mood became our mood; we all reveled in a common abundance and in each other's well-being. I cannot recall feeling, in a settled country, a like sensitivity to the mood of the land. (Leopold 1949, 146)

The Colorado River passed through the imagination of Aldo Leopold as he traveled the delta. The river and its delta then emerged in Leopold's writings as part of the inspiration for a new and founding "land ethic" in the environmental movement. The delta and its abundance gave Leopold an education in connection to the land—the land as a community that includes habitat, creatures, and humans.

That the landscape that inspired Leopold's epiphany could have been so damaged and degraded by dams on the river was a travesty to the visionary. Yet three decades after Leopold's death, the Ciénega has emerged from the wastewater of the Welton–Mohawk. I'm sure that now he would love to return to see the Ciénega de Santa Clara, the new "green lagoons."

In addition to the natural values of the delta—the birds and the endangered species like clapper rails—Leopold's long-ago visit to the green lagoons gives this area an historical significance that deserves commemoration in its own right. We should protect for the rails, yes. But we should also preserve it, if not in Leopold's name, then at least in some measure in his honor.

❦

Along the eastern side of the Ciénega, the delta butts up against the mesa of the Sonoran Desert. A geological fault line runs along the trough of the Gulf of California and the silted bottomlands of the desert delta. Through cracks in this still-active fault, an underlying aquifer bubbles to the surface in an uneven line of freshwater springs. Most are small watery oases. A couple are big enough to make tree-ringed ponds, trickling down a gradual slope into the Ciénega's eastern marshes.

These springs are joys—cool water in the desert heat, green islands in the brown sand, and refuges for wildlife. From the eastern edge of the mesa, you can get stunning views of the entire Mexican delta, especially at sunset. These springs lie just below the sandy bluffs. One of the most beautiful is called, fittingly, *La Flor del Desierto*, "The Flower of the Desert." South of the Welton–Mohawk Canal off

Mexican Highway 40, La Flor is a great place for birds. It's also where, in one survey in the late 1990s, biosphere reserve director José Campoy reported seeing two birds that he "suspected to be black rails" (Priest and Campoy 1998, 8).

Because California has lost so much of its wetlands (more than any state in the United States), the California black rail is endangered in the state. The bird is not listed in Mexico—not because the rail is safe, but because it is so rare, so unstudied, that not enough data has been collected on it. It's fair to say that black rails are among the continent's least known birds. They are the smallest North American rail—barely bigger than a large sparrow—and are not only fewer in number, but they are also the most secretive.

In our own search for a black rail, Juan Butrón and I have come to the pond at La Flor del Desierto. For all the decades that I have been bird-watching, I have wanted to find a black rail, the only rail species in North America that I have not seen. I have hiked huge marshes in my time, sometimes with only a flashlight at night, when certain species are most active, looking for the elusive rails.

Juan looks forward to finding a black rail as well. He has not seen one either and he is eager to see new birds in the Ciénega. Juan's *ejido* still owns property that extends this far to the east. He shows me the property line. Driving through dry sand along the edge of the property line, we pass through a stand of *tornillo* or screw bean mesquite, one of the few places that so much screw bean mesquite can still be found in the delta, Juan tells me.

The spring is a clear pond surrounded by trees and filled with lush cattails and reeds. A black-and-white warbler, returning on migration, works a large cottonwood. Crossing the pond is a makeshift bridge of wooden planks that leads to a trail that drops into the depression of the delta. Mudflats surround the marsh below, the beginnings of the vast tidal regions of the delta. The canal cuts through the mud and shallow waters. Sanderlings and western sandpipers, red phalaropes and snowy plovers dart and spin in the water. The canal carries the fresh *aqua dulce* of the spring out into the reedy marsh of the Ciénega.

Black rails have strict habitat requirements. They need shallow water with gently sloping inclines along the edges. The black rails are suspected of inhabiting the pond on the small bluff above the marsh. Seeing them, though, is far more difficult than seeing a Yuma clapper rail. If you spot one, you are lucky. If you get anything more than a fleeting glimpse of a bird flying away, you have been blessed.

Osvel Hinojosa has been eager to collect information on these birds as we did with the census for the Yuma clapper rail, especially because the black rails are so rare and probably endangered. According to the literature, black rails were first reported along the lower Colorado River only in 1969.

The California subspecies of the black rail is the same one that occurs in Mexico. It has been sighted in only three places in Mexico. If it breeds in La Flor del Desierto or the Ciénega proper, the delta will be the only breeding place for this rare species anywhere in Mexico. But Osvel still has not found the nest of a Yuma clapper rail in the delta, much less a black rail, an indication of just how difficult finding rail nests is and how much work remains to be done on the birds in the delta.

Juan and I arrive at La Flor del Desierto in the late afternoon and walk down the trail along the canal into the marsh below. We play the tape of the black rail's call, a series of high *kee* sounds.

At one point, a small dark rail spooks at the base of reeds on the other side of the canal. I get a good look at it, disappointed that it is a sora rail. Out by the reeds, we can hear Yuma clapper rails singing. A Virginia rail skitters out of the cattails and stands fully exposed some distance away, racing back and forth, in and out of the cover.

We hike back up along the canal to the spring with the big cottonwood. I walk in front, crossing the planked bridge. As I walk down a wobbly board toward the muddy bank, a small dark bird flushes. It flies on rapid wing beats low across the pond, long feet trailing, and disappears into the reeds on the far side. The bird is small, dark, and has grayish wings. As it flies, it calls out in alarm. The call was unmistakable—*kee-kee-kee-kee.*

I know the call from the many tapes that Juan and I have played in our rail survey. I also know the call from the two black rails that answered us near the mouth of the Welton-Mohawk Canal in the Ciénega. It is the call of a black rail.

Juan and I search and wait and watch for another three hours. But the little bird—certainly a black rail—never reemerges. Two months later, however, in this exact location, Juan will see an unmistakable black rail cross the pond by walking up the planks and across the bridge.

As evening comes on, we give up the search, tantalized and keen for a better look, but still exhilarated. I have glimpsed a black rail for about one one-hundredth of a second, as it flew away and behind a bush. It may have been only a brief encounter after some twenty-five years of desire and searching, but it was a happy glimpse. Wanting to savor the end of the afternoon, we return along the canal down into the marsh, where from the sweeping and open flats we can watch the sun set over the delta.

We walk down to the edge of the huge marsh. The golden desert light renders the delta superbly beautiful—water and mountains and a transcendentally luminous evening light. We stand in the salt grass and the salicornia on the edge of the marsh and look to the west. The mountains of the Sierra Cucapá define the western border of the delta in the far distance where the sun will soon set. The dark silhouettes of avocets and stilts reflect off the waters of the marsh. In the distance, a flock of some forty or fifty white pelicans rests near reeds.

I ask Juan what his dreams for the marsh are. He tells me that what he really wants is to have some water dedicated to the Ciénega. "A little water," he says, "that's all we need. A little water dedicated to ecological purposes." *Con fines ecologicos.*

This is the rallying cry of the delta—water for ecological purposes. The trick is to find the legal mechanisms to make that possible. The justice of Juan's dream seems undeniable. Water is the life of the marsh, one of the great wetlands in the American West. After a century of growth and exploitation, it is the delta's time.

What is required is a change in the conception of water in the West. We need a vision—one that sees water used for the environment, even a little water, as a valued use. This would be a new value in a region that has considered even a drop of water reaching the Gulf of California as wasted. Such a shift in perception does not mean a radical new rechanneling of water flows. It does mean putting at least some water into our budgets for places and creatures that evolved in this region long before heavy human exploitation of the river began.

"I like this form of life," Juan says about his life beside the Ciénega. *Esta manera de vivir.* Juan does not have much, and he understands that, especially in comparison to the material standards of the United States. But he knows and values what he has.

"We have to take care of nature, protect it," he says. "And live in a sustainable way for our families, so our kids can keep on living well too. It's a wonder. You know my son, he has lived here since he was tiny and he knows this reserve so very well."

Juan could not have articulated the human values of the Ciénega more perfectly.

The Ciénega is also a singular example of the emergence of something new in our understanding of nature. The spring water from La Flor del Desierto feeds the Ciénega with natural water from the aquifer beneath the delta. But most of this huge marsh is formed by wastewaters pumped some fifty miles south from the United States. In one sense, this is an artificial habitat, completely encircled by management decisions. It may be wild, but it is not natural.

A large debate has emerged in theoretical circles about whether we can ever have access to a pristine nature, or whether nature is always in some sense constructed and therefore artificial in our human experience. Perhaps, because nature is no longer pure, but is everywhere stamped by human enterprise, nature itself is dead. All nature, in this view, is a kind of artificial "second nature."

The Ciénega defies our customary categories of "artificial" and "natural." Along the Colorado River, the two categories are completely entwined in each other. The thousands of endangered Yuma clapper rails in the Ciénega, and the few very rare black rails, do not care whether this habitat is pristine or artificial. Water has returned. They have thrived in the "wastewater." The Ciénega de

Santa Clara is an example of a unique wilderness, a marsh that shows the potential for a recovery in the delta and that confuses our traditional categories of nature and culture. The delta is tutoring us in a new kind of restoration of nature. To rejuvenate the delta, we need only one basic element. As Ed Glenn put it succinctly for me one time, "All we need to do is give water. The plants and animals will come."

Juan looks out over the marsh, its line of cattails and reeds defining the middle distance. At our feet is the water; along the horizon are the mountains. The sun vanishes behind the jagged peaks. Wispy cirrus and altocumulus clouds turn pinkish, reflecting softly off the marshy water. The clouds grow brighter, an incandescent passion of pink, shading through orange into a burning yellow above the mountains.

Juan himself is the best example that, in the Ciénega, you can't separate the human from the natural. These are his *terenos familiares,* as he puts it. The emotional force of that phrase is impossible to overstate—family lands, the lands he knows. The rich light glows on his face. I look into his eyes and see the entire world of Ciénega.

*La Flor del Desierto blooms into an exquisite sunset that sweeps over the delta and the spectacular Ciénega.*

# THREE

# An Arrested, Shrinking Delta

PROPOSED THREATENED SPECIES: Flat-tailed horned lizard
PLACE: El Golfo Badlands

You KNOW IT the minute you enter this part of the delta. El Golfo Badlands is not simply unvisited, it's positively avoided. Some off-road vehicle enthusiasts, driving down in big rigs along Mexican Highway 40, have discovered the sandy beaches and dunes along this stretch beside the northern Gulf of California. Daring nearly vertical hills, they spit sand and snarl through the silence. But the off-road vehicles (ORVs) hug the coast, never far from the town. Hike only a mile through the dry washes and you will leave them behind—no more engines, no more tires tearing up the sand and silence.

Even the fearless Jesuit missionary, Father Eusebio Kino, failed to cross this section of desert. In 1702, Kino pioneered the land route through Mexico as far as the Colorado River, following the San Pedro and Gila Rivers through what is now Arizona. When he got to the big river of the north, he gave it the name that stuck: the Colorado River. Venturing downriver with throngs of adoring Native Americans, he was the first Spanish missionary in the delta. To return, he tried a more direct route from the northern Gulf east through the Gran Desierto to Sonóita. Few deserts daunted him, but he gave up here. Kino's great biographer Herbert Bolton in *The Rim of Christendom* describes the failed attempt as one of "the hardest and most dangerous marches imaginable." "Few white men," concludes Bolton, "have ever traversed the route they attempted" (Bolton 1936, 482–3).

Mexican prospectors also skirted this part of the desert as they rushed to the California gold fields in the mid-nineteenth century, instead swinging north along a course that was still brutal enough to be called a *jornada del muerto*—a death march.

◄ *The rippled sands and sedimentary mesas give the desert delta a sensuous beauty.*

▲ *Well-adapted for desert life, this harmless little lizard looks fierce as a dinosaur. It survives by eating ants and hiding from predators by burrowing into loose sand.*

I am here to prospect as well, but not for gold and definitely, in this forsaken area, not for souls to save. I am here with geologists and paleontologists. We are looking for old bones and we have the area completely to ourselves.

We are in the "Badlands," *Tierras Malas*. They begin just north of the fishing pueblo of El Golfo de Santa Clara and extend about twenty miles south along the edge of the Gulf of California. They stretch inland for miles, covering about one hundred square miles of dangerous and deadly desert. They are part of the mesa of uplifted land that stretches south from Yuma, five hundred feet high in some places. This rolling landscape of hills and valleys holds the windblown remains of prehistoric seas and ancient riverbeds of the Colorado River.

The landscape is severe, punishing the senses in its ascetic spareness and heat, especially by midday, when the vertical sun is unrelenting. Still, there is something deeply sensuous about this area. The curve and contour of the prehistoric hills have a seductive beauty. In the early morning the light and shadows play erotically with each other on the flesh-colored, wind-rippled sands.

The people I am with are modern desert rats. They are good at dealing with this desert. They have been among the few people who have made a point of coming here. Fred Croxen III is a slight man with an enormous capacity for endurance. A geologist for the Bureau of Reclamation's offices in Yuma, he has been coming to the delta since the early 1970s. He conducted some of the early studies of the Ciénega de Santa Clara. He is also a geology professor at Arizona Western College. As a child he loved finding fossils. In 1988, his long-simmering passion for paleontology burst into a full-blown obsession here in the delta Badlands. Ever since, he has been bringing his geology classes down to the Badlands. I am with him on a field trip for his class.

With us is Chris Shaw, Collections Manager of the George C. Page Museum in Los Angeles, better known as the La Brea Tar Pits. In 1981, Chris did his master's thesis on the fossils of this El Golfo region. It was a landmark study, revealing an unsuspected paleo-ecology of the area. He literally camped in the desert, crawling on his hands and knees in the sand to look for bones. It turns out that this part of the delta is extraordinarily rich in Pleistocene fossils—the period of time stretching back about 1.6 million years.

Chris and Fred are now exploring the area together to create a picture of the Pleistocene ecology of the delta. From the more than fifty-two species they have identified from the fossils discovered here, they know that the Badlands was once a very different and extremely diverse world. The creatures include some of the most exotic to have been found on the continent.

"The mesa here is sedimentary from the Colorado River," Fred says as he prepares to head into the desert area to look for fossils. We will form several small groups and fan out through the draws and arroyos. "It was lifted up geologically, which means it's been exposed for thousands of years. It's a rich area.

Maybe not as rich as La Brea, but great. It's yielded some important discoveries."

The land is like a geological bulletin laid down long ago. Winds scour the bare hills. The occasional rains wash away the mud. Slowly, over periods of thousands of years, the bones are exposed to view. They may be lying on the surface, partially visible. Or they may tumble out of the hills and slide down in small bony avalanches.

Our strategy is to walk the arroyos and abraded hills and watch for bones. It takes a trained eye to begin to spot them. Once you get the idea, searching is enormously fun. On almost every field trip, Fred and his students have produced new finds, significant contributions to the understanding of the evolution of life on this part of the continent. There is nothing more exciting than finding a new species. Such a discovery is the Holy Grail of fossil hunting, for amateurs and professionals alike.

Chris is especially hoping to find the greatest prize of the area. He is almost certain that saber-toothed cats once roamed the delta. He has found fossils he believes are *Homotherium*, in the saber-toothed family, but nothing definitive yet. The creature hovers in the talk and the imagination of everyone on the trip as we head out.

As we leave the spot where we have parked the pick-ups, I follow Fred. We walk and chat while he watches the ground. We have gone less than one hundred yards when he stops. He points down to a patch of ground. Sand. Some small gravel. A creosote bush.

Fred bends down and points. Barely exposed is a knucklebone.

"That's a camel toe," Fred says without missing a beat. Fred has the natural patience of an excellent teacher, happy to share what he knows. This bone, he says, is perhaps a million years old. It dates to when camels were one of the most abundant animals in the delta, at a time when the area was teeming with creatures. That camels were here suggests how different in climate and creatures the delta once was, before it became a desert.

It is miraculous to be confronted with a remnant of life that old. The bone is nondescript, but suddenly a whole new perspective on the landscape opens wide. Perhaps the most exciting part of finding a buried fossil is what it does to the imagination. No wonder that small children love prehistory, with its strange and unbelievable animals. Kneeling with Fred to inspect the bone, I feel the empty desert open to the imagination and come alive in my mind. The delta's geological history transforms from dead rocks and bones into a living process.

What became the Colorado River began laying down sediments in the delta as long as 4.5 million years ago, according to Fred. The Badlands, he says, represent the last 1 to 2 million years of the deposits.

The well-worn hills around us are part of a long geological process. Everything around us was created initially by the Colorado River. And not just the exposed Badlands, where the geological processes are very visible. The river created the entire delta, which is bigger than some states in the United States.

The Colorado River is not the biggest river in North America. It is only the seventh longest. The longest river is the Missouri, which is more than one thousand miles longer. The Colorado barely makes it into the top twenty-five rivers on the continent in terms of water volume. Compared to the Mississippi River, it is a trickle; the Mississippi carries nearly twenty-five times more water than the Colorado. But the Colorado carried a silt load unrivaled by any river in the world. It was a phenomenal, even legendary, burden of mud. The Mississippi River may have been much longer and carried much more water, but the Colorado River carried seventeen times the amount of silt. And in relation to the most mythic of rivers, the Colorado River carried ten times more silt than the Nile River in Africa.

The Colorado originates high in the Rocky Mountains 1,700 miles to the north. From Wyoming to Mexico, it drains 246,500 square miles of watershed, one-twelfth of the continental United States. It was famous for its wildness, roaring out of the Rockies in a crazy plunge unmatched by any other river in the United States. The ferocity of its natural flow is hard to imagine any more, though the narrative of John Wesley Powell's first descent through the Grand Canyon gives a picture of the former jaw-rattling and stomach-churning fall as the river tore southward. The river's power was such that it cut through the easily erodable Colorado Plateau and created the enormous western canyons that are so famous and are so much a part of the historical identity of the United States. The river—and all its geological ancestors—was one of the great, wild rivers in North America, capable of ripping out the biggest canyon in the entire world. Small wonder the river and its Grand Canyon are icons of America, almost characters in their own right in American history.

Some descriptions say that by the time the river emerged from the canyonlands it was worn and spent. Others characterize the river at this point as having glided slowly and with a leisurely, magnificent dignity through the deserts of what are now Arizona, California, and Mexico. At any rate, after its wild trip through the canyons the river was carrying a big chunk of the continent it had passed through. Its flow was a slurry of mud. "Colorado" means red, not just "colored." The waters were not just a muddy, dirty brown, but had a striking red cast due to the iron-rich soils on the plateau. Small stretches of the river still at times exhibit this color—the color of blood. The river as lifeblood and lifeline.

When it was running high in the spring and summer—and prior to being dammed—the river carried as much as one million tons of silt and sand past any single point in a single day. You will encounter widely varying numbers on the amount of silt the river carried. The definitive numbers come from the Bureau of Reclamation, which was smart enough to want to know as much as possible about the dragon it was fixing to tame with dams. So before the dams went in, the bureau put gauges in the river near Yuma to measure the silt carried in the water. In the years just before the dams, about 147 million tons of silt passed Yuma each year.

The most reliable modern discussion of the sedimentation of the delta and the Colorado River is found in the book by geologist Robert Wayne Thompson, *Tidal Sedimentation of the Colorado River Delta, Northwestern Gulf of California*. He estimates that the original sediment load of the river was about 20 percent sand, 80 percent clay and silt.

How much silt and sand is this? The volume is hard to grasp. But day after day, through millions of years, the Colorado River laid down its load in the delta. About 70 percent of everything carried by the river got dumped in the delta. To match its earth-making ability, you would have to back a double dump truck full of dirt up to the mesa at Yuma and unload it every six seconds, day in and day out. Or imagine routing the river through the brilliant new baseball stadium of the Seattle Mariners, Safeco Field. It holds 45,000 people. The river would fill the stadium with mud in about fifty minutes.

Over millions of years, the deposited sediments grew so deep that bedrock is nearly impossible to find any more in the delta. How deep are these layers of mud, sand, and gravel? The depth varies throughout the delta. The deepest point is at a spot in the "deltaic cone," the high spot just near the current international boundary that separates the Imperial Valley in the old Salton Sink from the Mexicali Valley. Geologists estimate the depth of sedimentation at this location to be about 6.4 kilometers—nearly twenty thousand feet. The deepest deposits go back to ancient Miocene seas, about 12.5 million years ago. So you would need to go a long way down to find bedrock, and then you would be approaching the earth's mantle, itself floating on magma.

Carried by tidal actions and currents, river sediments reach south into the top third of the Gulf of California and are found in the seabed. Sediments from the river were deposited almost as far north as Palm Springs and San Gorgonio Pass, cutting through the mountains toward the Pacific Coast and Los Angeles.

Before the dams were built, most of its silt rode the Colorado River down to the delta. Now all that silt is trapped behind the dams. The *colorado* is now the *clarado*, the "cleared" river, at least in the delta. Below some dams the river can still run very red after storms, carrying silts that will pile up behind the next dams and, inevitably, fill up the reservoirs. In this way, the river may get its revenge. If so, it will not be our problem. It will be our grandchildren's problem. Or their grandchildren's. We have bequeathed them a problem so large that

*The "Badlands of El Golfo" reveal the desert delta's greatest geological secret: the soils that the Colorado River carved out of the Grand Canyon were carried south and deposited here in the delta. Delta and canyon are two halves of a single geological whole.*

historians and geologists fear it could bring down the civilization we have created out of these dams in the desert.

In sum, the river built the delta. It was an ongoing process. The sediment deposits were a nutrient-rich, fertile source of perpetual renewal for the delta, and when we built the dams, the delta quit growing. All of the churned-up silt was stopped well short of the delta. The dams have not simply stolen the river's water, they have stolen the delta itself.

As a result, something strange has happened in the delta. In the past, the delta was a dynamic geological area. It grew every year with the addition of new silt. Witnesses in the last century described watching parts of the delta extending farther out into the Gulf over time. That growth has now been arrested.

Now there is a kind of double geological action going on. Coupled with the disappearance of silt deposits, another process, a marine one, is making the delta shrink as well. With no more silts traveling down to the delta, waves of the Gulf

are erodong the delta and carrying it out into the sea. Geologists and hydrologists describe the delta as an "anti-delta," or "a negative estuary." These are bizarre epithets, suggesting an unreal geological inversion zone. The delta itself is endangered, vanishing. It is an arrested, even shrinking delta, cut off from its own geological history.

The apparent contradiction of this chapter's title—a delta at once arrested and shrinking—is geologically accurate, depicting the vise, which is stealing the delta from itself. This contradictory process has produced a surreal landscape as strange as any envisioned by Salvador Dalí, where timepieces melt across the sands and ants inherit the world.

Yet as I stood in the Badlands with Fred, imagining the delta at a time before the upriver dams, my imagination was most struck by one fact. All this land that spread out before us in scorched and scoured hills had once been a part of the Rocky Mountains. It had all come from what was once the United States. Americans visit the Grand Canyon and gape in awe. The delta of the Colorado River is where all that gouged-out earth was delivered.

The Grand Canyon and the delta are two parts of one original, geological whole. Mirror images of each other, they are two sides of what was once a single, living river. The dirt beneath us here in Mexico was once part of the earth in the United States—before these two countries and their unnatural boundaries existed.

From fossils like the camel toe that we found, Fred and Chris have constructed a picture of the Pleistocene world of the Colorado River delta that is nothing like the desert we are standing in. A million years ago, these heavily weathered hills would have been part of an almost tropical savanna. This camel would have lived amid a mixture of broad grasslands, trees, and woods. We are walking through a double world: the one visible to our eyes, and the one opening up from the fragments of bones in the soil.

"Water flowed right through here," Fred says, bagging the camel toe in plastic. "The climate was completely different. It had to be really nice. Fertile. Amazing. Even tropical."

Through the fossils, Fred and Chris have discovered that these Badlands in the delta were part of a savanna corridor running from South America into southwestern North America. From the Badlands fossils, Chris has determined that the sediments must have been laid down between 400,000 and 1.8 million years ago. The deposits can't be younger than 400,000 years because there are no bison remains. Bison are a marker creature; they are known to have crossed the land barriers into North America sometime after this boundary date.

All the mammals the team has discovered are from this broad time frame and are known to have lived in wet, tropical habitats. "This was a haven here," Fred goes on. "We know it was lush, 'cuz we've found bones from boa constrictors. We found a tortoise a lot like the ones in the Galapagos Islands. Chris identified bones of a giant anteater. It's the first record ever for an anteater this far north." The delta anteater is the only one ever found in North America. The fossils were found 1,300 miles farther north than any anteater yet recorded.

A wonderful list of creatures once roamed the prehistoric Colorado delta. Some of them will seem familiar, others are a shock. Many of the species survived for hundreds of thousands of years. The camels represent a number of species—horses, sloths, elephants, and large carnivores—that went extinct at the end of the Pleistocene, about 11,000 years ago. The postglacial climate change, and some human predation, probably did them in. Others went extinct very recently, but not because of any change in the river or the climate. They were driven to extinction by modern white human beings, who moved into the region only in the last century.

A wolf much like the modern red wolf roamed the grasslands here for hundreds of thousands of years. The red wolf was driven nearly to extinction in the southern United States in the last century by concerted predator control programs. There was also a strange hybrid, a "coyote wolf"—perhaps the progenitor of the red wolf, which has often been linked with coyotes.

Along with other predators, the wolves hunted both the mule deer and the white-tailed deer. The mule deer survived for hundreds of thousands of years in the watery delta, only extirpated from the region in the last fifty years, a much-regretted victim of the dams.

Black vultures shared the skies with a recently discovered extinct species of condor. This condor was very similar to the nearly extinct California condor,

*Dragonflies bring a prehistoric feel to the dry desert of the delta and likely flew through the former woodlands of this region when saber-toothed cats were here.*

which itself is a kind of prehistoric bird that once flew over the Grand Canyon and almost certainly over the river delta. Great horned owls hunted in the woodlands, finding an abundance of rodents, including gophers and two species of wood rats.

Raccoons worked the many waters of the ancient delta. A prehistoric species of beaver was abundant throughout woodlands and wetlands. Modern beavers were enormously common in the more recent delta. Now they are rare.

Badgers burrowed in the sedimentary soils, snarling from prehistory until now.

The spectacled bear, *Tremarctos*, left his remains in these desert hills.

More intimate and endearing, mud turtles settled into the sloughs, a slider pond turtle slid from logs in the backwaters, and a Colorado River toad was already spreading spawn in the delta a million years ago.

To a human being transported back in time, the world of the delta would seem astonishingly exotic in its tropical abundance. The landscape produced a faunal extravaganza, a kind of refuge for creatures and a breeding ground on a massive scale.

Amid this abundance the four species of camels might have looked vaguely familiar. But the huge Cuvier's gomphothere would have announced in no uncertain terms that this was a different delta. "It was a strange-looking mastodon," Fred says. "It had foot-long molars."

Whereas the contemporary delta is a southern extension of many northern habitats, during the Pliestocene the opposite was true. South America once reached up to North America and the two met in the delta. It was an exotic world of southern tropical creatures. "It was very interesting here," Fred says, "because there was a faunal exchange between the two continents."

A second elephant-like creature, the southern mammoth, shared the delta with the gomphothere.

Two species of tapirs, giant grazers in tropical forests, worked the woodlands along the Pleistocene river. Three species of ground sloths have been dug up in the delta.

Capybaras are an enormous rodent found in South America now. They were a prime food for jaguars in the delta swamplands.

A species of hyena, the American hunting hyena, was one of the more exotic creatures in the delta.

But the preeminent symbol of the glory of tropical habitats is the jaguar. Fossilized remains show that jaguars prowled the delta way back in the Pleistocene. The bones of a small, ocelot-sized cat, the "Rexroad cat," have also been found. And very recently, one of Fred's students discovered a mandible with teeth of an amazing creature—an American cheetah.

The jungle of swamps and woodlands would have suggested some of the great habitats of the world, with a biodiversity reminiscent of the great swamps of the Pantanal in Brazil. The delta might also have rivaled the Great Plains of North America, legendary for herds of grazers that stunned humans into awed silence. Among the great numbers of creatures in the delta were vast herds of grazing animals. Three species of antelopes kept company with camels. Two species of llamas were found this far north of the Andes of South America.

Three species of horses roamed the plains. These creatures went extinct and were only reintroduced to the continent when the Spaniards arrived at the beginning of historical time for this area.

"The herbivores are the really big players here," Fred says. "Camels, deer, and antelope are about 50 percent of all the finds. Another 33 percent are the odd-toed grazers—horses and tapirs."

One of the species of tapir was found by a person in the group only yesterday.

"Small creatures like fish and reptiles are discriminated against in the record," Fred says. What he wants to do now is a sophisticated chemical analysis of the herbivores' teeth to see if he can begin to identify the prehistoric plant community and define the climate.

Today's unearthed camel bone is like a seismic shock to the consciousness in a desert that shows no signs whatsoever to the uneducated eye that lakes and streams were once here. The rich fossil grounds around El Golfo are one of the amazing legacies of the delta—a partially exposed treasure trove rich in Pleistocene deposits of dried-up river mud.

This morning Fred hikes off to map a part of the Badlands beyond the next line of hills. Much of what he is currently doing is mapping potential fossil fields using the Global Positioning System (GPS). Soon he and a couple of students have disappeared.

I stay with Dr. David Sussman and his wife, Dolores. He is an anesthesiologist and both he and his wife are amateur paleontologists who have been prospecting in the delta for years. They teach me how to find fossils. As we round one old hill, Dolores spots a very yellow stone that turns out, on inspection, to be a flat bone. David looks it over carefully. Perhaps a horse, he thinks. We soon discover that the whole base of this hill is full of bones, all of them from a horse. We realize, too, that all of the bones are probably from the same horse. The animal no doubt died and was covered by sediments from higher on the hill. As the land was exposed and the river receded, the hill eroded. The horse came apart, and the disarticulated skeleton of the horse slid down the hill in bits and pieces until, thousands of years later, we stumble upon them. Only now the bones are not in a woodland but a desert. They lie among orange pebbles and slabs of sandstone. A few scraggly brittle brush are the only plants.

We bag the bones and take GPS readings of their exact positions. It is enormously fun, enormously rewarding.

We have agreed to meet back at the trucks at one o'clock. When we get there, everyone is thrilled to share their discoveries. On returning, Fred unfolds a blue table under the searing sun, and beneath broad-brimmed hats we lay out the bones amid Coke cans and potato chip bags.

Fred confirms that we have found a horse. Maureen has found another camel bone. Then another person, Randy Reisland, lays down a tiny bone. Chris Shaw from the La Brea Tar Pits gets excited. He looks it over carefully. That Randy could have spotted the tiny fragment of bone is amazing in itself. Chris's ability to identify it is equally impressive. He congratulates Randy for finding a new species of animal for the delta. Randy has found the jaw of a gopher. It's no saber-toothed cat, but it's a thrilling achievement.

We are all exhausted and sweaty from the heat but genuinely thrilled by the adventure. "We been rode hard and put down wet," one of the students laughs.

Lying on the table, the bones bear mute witness to a wonderful former world, rich and abundant, so different from the world around us at the blue table. These fragments deepen our view of, and our feelings for, these lands. The little pieces are planetary nostalgia, connecting us to the earth through grand flows of time and life.

The river is only one part of the geological history of the Colorado delta. The other part is the Gulf of California. The delta soils were laid down in the opening created by the great tectonic forces that tore open a huge gash in the earth's crust. You might imagine the delta at the convergence of two great geological processes. One is the river, grand conveyor of muds and silts. The other is the great inland trough into which the massive loads of sediments are dumped.

Think of the Colorado River as a narrative, a kind of novel that flows through time. The other geological story in the delta is an epic drama of the collision of sliding tectonic plates. The two stories almost cancel each other out, in a sense. While the crash of plates created dramatic geological features like seas and mountains, the river's huge sediment load was busy obliterating this topography with featureless deposits of primordial mud.

The magnificent uniqueness of the Colorado River's deltaic systems, geologically speaking, comes from the combination of the river's unprecedented silt load with one of the most remarkable inland seas in the world. According to one of the earliest geographers to survey the region, the Gulf and its delta are of "even greater magnitude than the now famous prototype" in the Nile and Mediterranean Sea. (Kniffen 1932, 156)

The river flows into what is now the head of the Gulf of California. Geologists are still putting together the exact story of the creation of this long trough between two continental plates. If we think of the area as a human body, the Baja peninsula is an arm, Mexico proper is the body, and the river empties into the armpit.

The Gulf began to emerge between 20 million and 15 million years ago in the aftermath of the Pacific and the North American plates' collision. These tectonic plates met in a slow gnash and grind, moving portions of the earth's crust. Parts of the continent were uplifted, volcanoes on what are now the peninsula and the coast of Mexico sprouted up, and land masses split open.

Out of this primordial violence, the Pacific plate ripped off a chunk of North America more than 10 million years ago in the Miocene. The Pacific plate raced off with the torn body part of North America. The Baja Peninsula and southern California became part of the Pacific plate and, by the time standards of geological measurement, they might as well be sprinting north and west, some inches per millennium. What is more, they continue to move even now.

As the Baja tore off from the continent, the earth in between sank and left a long, heavily faulted trough. Geologists describe it as a rift-and-fault system. The faults lie in offset parallel series. Like long geological stretch marks, they are at once a sign of the expansion in the bedrock and sites of incredible geological activity. The head of the Gulf of California is still one of the most active geological places in North America. The San Andreas fault is only the best-known fault because it cuts all the way up from underneath the current delta at the north of the Gulf into the heavily populated regions of southern California. Along its sheered line, the Pacific plate continues to rip off, with many other faults as evidence of the underlying events. Under the Salton Basin, near the California border with Arizona, the San Andreas sidesteps to a parallel fault, the Imperial fault, which in turn sheers into the Cerro Prieto fault, and so on down under the enclosed sea. A small chunk of the ocean floor actually still lies beneath the Salton Sea in California.

Cerro Prieto, or "Black Butte," looms in the heart of the Mexican delta and is an extinct volcano. Earthquakes, tremors, still active and gurgling mud volcanoes, boiling geothermal ponds whose temperatures reach as much as 350° C, and freshwater springs in the hot desert—these are only a few signs of the deep geological drama taking place under foot, day by day. Baja California itself continues to migrate north, having moved some three hundred kilometers in the last 10 million years, and continues to rip part of the state of California from North America along with it.

Between 8 million and 5 million years ago, the Pacific Ocean surged into the growing basin, with new crusts forming on the spreading ocean floor. The seas rushed into the long trough inside the peninsula to meet the waters of the Colorado River.

The current geological configuration of the delta is probably pretty recent, the tectonic features brought to their present shape by recent "considerable uplift." The long mesa to the east of the delta, running south from Yuma and down into Sonora, passes along the upper northeast edge of the Gulf. It is the product of both doming between fault lines—a "sheer zone"—that raised the land up above the plain of the Gulf, and of changing world sea levels during past glaciations. Once this mesa bulged out of the low-lying landscape, about eight thousand or so years ago, the river began emptying near Yuma, into the area we now think of as the delta.

Running high with spring floods, the Colorado River roared out of the mountains, made a right turn to the south near what is now Yuma, and found its delta lying before it like a huge catchment basin for its silts. Bulimic and engorged with dirt, the river dropped its load on the level plains, cutting its own channels as it crossed the trough. The silt piled up, filled the channel, and the river raced off somewhere else, cutting new channels. Sometimes its own silt would block the river's southern flow, and the ancient Colorado would tear off to the north, dumping all its silt and its waters into a huge landlocked prehistoric lake, Lake Cahuila—precursor to the modern Salton Sea. In the last few thousand years, this lake has filled and dried up several times. Calcium and shells can be found on the surrounding mountains like a huge bathtub ring.

Evidence suggests that this ancient lake may have been filled even as the very first Spanish explorers were entering the delta in the sixteenth century. None saw it. None went this far into the desert. The first Spaniards were at the delta by 1539 and Lake Cahuila is thought to have evaporated by about 1580.

Between the northern extent of the delta, in California, and the southern delta now in Mexico, the deltaic soils form a small dome, barely thirty feet above sea level. Old channels of the ancient river cut through this deltaic cone, cutting a route into the ancient sea—the route for the river when it ran at its schizophrenic best. The Bee, the New, the Alamo, the Paredones—all these rivers and washes provided outlets for the supercharged river when it wanted a new route through the delta. The result was a maze of meanders and lakes and natural levees as the upper trough filled in and the river whipped back and forth through its delta. Below the cone that divides the delta in two, the river created a broad alluvial plain and huge flatlands of mud.

Two places in the delta still have not filled in completely. The Salton Sink is about 280 feet below sea level. The Laguna Salada in Mexico's delta is about 10 feet below sea level.

The Badlands, where the mastodons and anteaters thrived, rose upward about 100,000 years ago into terraced sandy hills. In deep time, this land lay beneath Pliocene seas. The river came with its massive sediment load and forced the retreat of the early Gulf of California by creating a "sediment plug" near its apex. When the seas receded, the river passed through the region, leaving deep deposits of alluvial soils. Into this region, the Pleistocene animals moved happily, prospering in their tropical refuge of wetlands, grasslands, rivers, and sea. When the land domed up in a "sheer zone" between faults—part of the same process that created the present configuration on the eastern side of the delta—the sediments were left high and dry.

Both the river and time itself seemed to abandon the Badlands, leaving it to lizards and brittle brush and the winnowing winds.

Measured against the long perspective of geological time and processes, human ambition in the desert delta seems small and insignificant. One of the most beautiful writers to love the delta was John Van Dyke, who entered the desert just at the time the first white settlers were beginning to divert the river and dream of agricultural empires in the river-rich soils and hot sun. In 1898, this art professor from Rutgers University left his post and began three years of solitary wandering in the great American and Mexican deserts. His book, *The Desert*, is one of the most gorgeous and compelling tributes to this region that has been written, a classic of sensitive observation and reflection that deserves to be much more widely known.

The first desert Van Dyke enters in *The Desert* is the one along the Colorado River stretching down into the delta. He describes learning a new way of seeing in these deserts. He calls it "sensuous seeing," trained by the opalescent light of the desert. In the chapter, "Down in the Bottom of the Bowl," he writes about the below-sea-level desert in the Salton Sink of the delta: "The opalescent mirage will waver skyward on wings of light, serene in its solitude, though no human eye nor human tongue speaks its loveliness" (Van Dyke [1903], 1980, 62).

These spaces, he writes, are crucial to protect. They are of enormous value to American civilization, though few have realized it. "They are the breathing spaces of the West," he writes movingly, "and should be preserved forever" (Van Dyke [1903], 1980, 59).

But they were not.

The dams have changed the river and enabled farmers and cities to move in. Even as he records his profound love for the lonely and lovely landscape of the desert delta, Van Dyke prophesies the coming of its devastation. It will come, he says, at the hands of what he calls "the practical men"—the engineers, farmers, and city fathers. These "practical men," he warns, will "cut the throat of beauty" and "flay the fair face of these United States" (Van Dyke 60, 61).

Within a generation, the river no longer flowed into the delta. Van Dyke moralized in an almost biblical way about the building of a human empire in the desert and the delta: "Nothing human is of long duration," he wrote. "Men and their deeds are obliterated, the race itself fades; but Nature goes on calmly with her projects . . . ." (Van Dyke 62).

The revenge of the river has not yet come. We are still trying to prolong our Faustian bargain in the desert, all the river's water in exchange for a few generations of time. The consequence is that the delta is an arrested geological system. Evolutionary and geological processes that were ongoing for millions of years, connecting the present to the past, have been frozen. It took more than several million years to create the delta. It took human beings less than one hundred to destroy it.

Several months later, I am back in the desert Badlands with geologist Fred Croxen. Maureen is here too, as is Robert Predmore. We are headed to an even more remote part of the Badlands, behind Punto Machorro. We cut across the beaches south of El Golfo at early morning low tide. A quick left turn between huge dunes of sand takes us into the vast drift of the Sonoran Desert just off the sea.

Robert is athletic looking. He works for the city of Holtville, California, as a city groundskeeper. He is accustomed to being outside in the desert heat in all

*The flat-tailed horned lizard was once popular and common in the desert delta on both sides of the border. Exploding human populations in the Southwest leave it looking out upon its vanishing desert habitats.*

seasons. Fred says that Robert is great at finding bones. Robert has found many of the new species in the Badlands.

"He's got radar in his eyes," says Fred. It was Robert, for example, who discovered the beautiful mandible of the extinct and spectacular American cheetah. Robert is more modest. He says he just likes "contributing to science."

On this field trip, he was to make another major new contribution to El Golfo paleontology.

About a mile inland, we leave the pick-up and start walking. We are following a dry wash that cuts through the middle of a grand set of hills. Again, we barely get away from the pick-up when Robert and Fred stop dead in their tracks.

We're looking for prehistoric creatures that are extinct. But I have told everyone to help me find what looks for all intents and purposes like a relict creature from prehistory; not an extinct creature, but a modern endangered creature. In addition to looking for bones, I want to find a flat-tailed horned lizard.

That's what they have spotted. Maureen and I rush over. The lizard is sitting on a shelf of sandstone looking out into the desert. It must be one of the icons of the West. It is surely one of the quintessential creatures of the deserts of the Southwest. A generation or two ago, these were deeply beloved by people living in the desert. Now front lawns and golf courses have taken over and few people see them regularly any more.

But still they are famous, affectionately known as "horny toads," because their fat and circular bodies look like a toad's more than a lizard's. Their genus name is *Phrynosoma*, or literally "toad-body." Few living creatures evoke the prehistoric ages of extinct creatures more beautifully. Flat-tailed horned lizards may be small—barely four inches from tip to tail—but they are unique in their armored saurian appearance, looking like some diminutive stegasaurus in spikes and plate armor. It is a desert fantasy, but it is a powerful one. In the 1960 film, *The Lost World,* the directors magnified tiny horned lizards hundreds of time to create "terrifying" images of prehistoric dinosaurs.

The small creatures are difficult to see because they blend in so well with the bleached-out sand and rocky *pedregal* of the desert. Despite their horny armor, their major defense is not ferocity. They have flat bodies. Their stomachs are rounded and almost comically flattened like dollar pancakes. They have a tiny tail, also flat as its name suggests, and they are arrayed with a fortification of spikes and horns, making a very prickly meal for any predator. It is hard to imagine a creature that could look at once more heavily armored and more vulnerable than this lizard.

There are sixteen species of horned lizards in the deserts of North America, eight of them in the United States, fifteen in Mexico. Many of them are famous for being able to spit blood from their eyes—a strategy that must surprise and temporarily confuse a predator. This is not the flat-tailed horned lizard's strategy. It depends entirely on camouflaging itself against the sand.

This flat-tailed horned lizard first tries to freeze. When we crouch close, it realizes it has been spotted and runs away. But not far. After about five feet, it freezes again, does a quick saurian shimmy, and wriggles into the sand. Nearly covered, it all but disappears in a sandy burrow. It stays frozen, partially buried in the sand. Its head still sticks out, though, and is flat, as if for head butting. It is covered in scales, black and beige and a faint sunset orange. Around its head is a spectacular crown of spikes. Between its camouflage and its spikes, the tiny lizard is not so much formidable as difficult to find and swallow—for burrowing owls or coyotes.

On my knees I get eye-to-eye with it. It has black beady eyes, looking back at me. It makes a fierce show, but the docile horny toad is all bluff. And it is utterly harmless. In fact, it helps humans, since its main food consists of ants, especially harvester ants. Children can pick these lizards up; since the creatures don't really flee, they are easy to catch. As long as we do not touch it, this lizard simply holds its partially hidden position.

While I study the lizard, Fred takes a GPS reading. GPS, or the Global Positioning System, gives exact locations for any place on earth, using small handheld devices that communicate with satellites. These devices are amazing in the field, crucial now for mapping of all sorts. Our exact place is 36 meters above sea level, 31° 39'53.76' latitude, 114° 25'57.79' longitude. Fred will give the reading to the biosphere reserve, which has begun the first study of the flat-tailed horned lizard in Mexico. Researchers want to develop population numbers, and they want to do a genetic study to see whether the lizards in Mexico are the same as the ones in the United States.

In the United States, the flat-tailed horned lizard has caused a major stir. It lives in the most severe desert habitat of any species of horned lizard. That has not, however, protected it from serious problems. Once, the flat-tailed horned lizard was much more common in the United States, with its range extending along the deserts of the Colorado River, along the Salton Sea down into the Sonoran Desert.

By 1993, scientists realized that the lizard's numbers were in decline. The main cause has been the unprecedented growth in the portion of its range in Arizona and California. This part of the American Southwest is among the fastest growing parts of the whole country. It is a familiar story—habitat has been vanishing as housing developments, golf courses grow, and off-road-vehicle courses grow. ORVs pose a terrible danger to the lizards. Since the lizards don't run, but try to flatten themselves into the sand, they are easily squished under oncoming tires.

The horny toad was proposed for listing under the Endangered Species Act in 1993. ORV and other groups protested. In 1997, the U.S. Fish and Wildlife

Service withdrew its listing proposal, claiming that evidence of a decline in the lizard's population was inconclusive. Instead, the agency proposed a voluntary conservation plan among various federal and state agencies, which included the designation of "management areas."

One of the biologists who has done the most work in censusing for flat-tailed horned lizards, Kevin Young of Utah State University, worries that the population of the species faces enormous pressures from the growth in the region. Defenders of Wildlife and other environmental groups took the Department of the Interior to court over the flat-tailed horned lizard. And won.

In July 2001, a unanimous three-judge panel for the 9th Circuit of the U.S. Court of Appeals found that the Secretary of the Interior had "relied on an improper standard and failed to consider important factors," granting a ruling in favor of Defenders of Wildlife. As a result, the Fish and Wildlife Service has once again proposed listing the flat-tailed horned lizard as a "threatened species."

Bill Snape Vice President of Law and Litigation for Defenders of Wildlife, hopes that this time the species will in fact be listed. "We have studies to show that habitat loss in the region is much worse than the agencies are acknowledging. Plus reptiles are notoriously hard to count. The estimates are pie-in-the-sky," he says. "We need the protection powers given by the Endangered Species Act."

In Mexico, the flat-tailed horned lizard population is thought to be in pretty good shape. But it's only a guess. Like so much in the delta, no one really knows—which is why the study is being conducted. During this day we found a total of three flat-tailed horned lizards in the Punto Machoro area of El Golfo.

When we leave the lizard, we go at most another one hundred yards and find another one. Fred and Robert take a reading and then leave. Maureen and I stay with the lizard. It sits on a ledge of sandstone, with the immensity of the desert in full view beyond it. I don't touch it, don't scare it away. I enjoy the huge view, receding beyond hills into a desert that stretches for hundreds of miles. It is a view, seen with the lizard, that stretches back in time as well. It is a fantasy, of course, but one I enjoy—trying to imagine the desert as this flat-tailed horned lizard sees it.

After a long morning, we gather again at the pick-up. Robert—the man with "radar in his eyes"—comes back with one of the most startling discoveries ever made in the area. He has found the first definitive jaw of a saber-toothed cat in the delta. Despite suspicions, there had been no conclusive evidence of saber-toothed cats in the delta—until Robert Predmore returned from a morning hike with one of its bones.

The flat-tailed horned lizard and the saber-toothed cat are two versions of the same phenomenon in the delta—the disappearance of species. Seeing the contrast between them is unavoidable. We normally think of the Pleistocene as the great period of mass extinction, yet we are now living in a time with mass extinctions greater than anything that took place in the Pleistocene.

It is important to note that the epidemic of modern extinctions is not the same as the extinction of the dinosaurs or the disappearance of saber-toothed cats. Endangered species are a uniquely modern phenomenon and from a wholly new category, one we humans invented. They are really a part of our culture— they are the totemic emblems of modern western culture. They are certainly not accidents, but rather the direct results of our way of relating to nature and our way of viewing creatures. Both—nature and creatures—are treated as inferior. Most troubling, endangered species are our favorite kind of animal. We like animals *because* they are endangered.

It seems to be the way we want animals—not exactly extinct, but reduced in numbers, heavily managed, living under our auspices. We designate management areas and interagency management groups. They become an industry in their own right, a part not of nature's economy, but of our own.

It is often said that extinction is a normal part of nature, now as always. But this latest wave of mass extinction is different from the earlier ones. We are living during the "sixth extinction," as E. O. Wilson explains in his superb overview of conservation and biology, *The Diversity of Life*. In four of the five early periods of mass extinction, "long-term climate change" was responsible. The plants and animals had long periods of time to evolve and adjust. Not so in modern endangerment and extinction. These processes are now caused by humans, characterized by rapid change in environments and the terrible violence done to nature.

These extinctions, Wilson writes, "should give pause to anyone who believes that what *Homo sapiens* destroys, nature will redeem. Maybe so, but not within any length of time that has meaning for contemporary humanity" (Wilson 1994, 31). He concludes his book with this sobering thought:

> If there is danger in the human trajectory, it is not so much in the survival of our own species as in the fulfillment of the ultimate irony of our organic evolution: in the instant of achieving understanding through the mind of man, life has doomed its most beautiful creations. And thus humanity closes the door to the past. (344)

As we leave the grand and vacant hills that lie back from the sea, carrying our saber-toothed treasure, we pass a fleet of ORVs. They race up and down one particular hill. Several others speed along the sandy beach. They are cocooned inside their own noise. I can't help wondering how many flat-tailed horned lizards they will smash today.

Beyond stretch the uplifted remains of a prehistoric delta, sweltering and silent, full of its living and dead relics—all of them partially revealed mysteries.

FOUR

# Super Fish

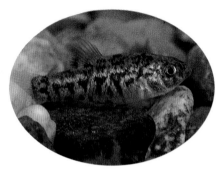

ENDANGERED SPECIES: Desert pupfish
PLACE: Cerro Prieto Lagoons

W̲ᴇ sᴛᴇᴘ ᴏᴜᴛ of the air-conditioned cab of the pick-up and slam into the heat. It's not merely climate. It's the only reality. The great desert writer Joseph Wood Krutch wrote in *Voice of the Desert* that to really experience the desert, you have to know it in the summer. During that season, it's most completely itself.

The desert delta, then, has declared itself.

It is July 24, late in the afternoon. The radio broadcast coming down from Yuma says the summer temperature has topped 115 degrees Fahrenheit. "That means it's probably hotter here," Jaqueline García-Hernandez says matter-of-factly.

Jaqueline García is driving the truck. She is a bright 26-year-old graduate student in the College of Agriculture, University of Arizona. She grew up in "D.F.," Distrito Federal, Mexico City. She studied as an undergraduate at Instituto Tecnologico de Estudios Superiores de Monterrey, in Guaymas, on the eastern coast of the Gulf of California. We have driven to the delta from Tucson. She is here to collect data for her doctoral thesis on contamination in the delta. On this trip, we are going to find little freshwater fish in the delta so Jaqueline can test for selenium poisoning.

We have just pulled up to one of the hot springs in the Planta Geothermica de Electricidad, the geothermal electrical plant, at Cerro Prieto south of Mexicali. These springs bubble up from volcanic activity below ground. With us is a plant engineer.

"*Chingo calor*," laughs the engineer, as we get ready to get to work. He is a bit surprised when I laugh too at his Mexican idiom. F****** heat.

◄ *At the height of summer the sun turns the desert delta molten in the morning. Temperatures will soar past 120° F near this spring in the desert, where pupfish survive.*

▲ *The desert pupfish is small but tough. Freshwater fish in the North American West have a rate of extinction and endangerment that rivals that of tropical rain forests. This little fish is on the brink of extinction.*

I ask him how he survives the heat. "In the morning we're hot," he answers in Spanish. "In the afternoon we catch on fire." *Nos quema.*

Jaqueline grabs her equipment in the back of the pick-up—thermometers, pH testers, salinity measuring tools, and a fine-mesh, long-handled fishing net. We have come to these hot springs because they have a large population of the last endemic freshwater fish in the delta. We lean over the pond near an intake pump. There, swimming near the rocks, is a small school of tiny fish, most about an inch and a half in length. They are very endangered.

This is the last place on earth—the industrial landscape of a desert geothermal plant—that I would have expected to find an endangered species. They are called desert pupfish (*Cyprinodon macularius*). This species is one of the preeminent endangered species, not simply in the delta, but on the entire lower Colorado River. The desert pupfish is found in the Ciénega and in several of the freshwater springs along the eastern margin of the delta. But we have come here because Jaqueline needs to sample this part of the delta for toxicity levels. She takes one of her long-handled nets from the rear of the truck. It has bright blue mesh. Leaning over the water, she stabs like a heron. No luck. The pupfish scatter.

One of the worst epidemics of extinction and endangerment in the world is taking place in this part of the North American continent—the arid Southwest along the Colorado River. The pattern of extinction and loss rivals anything seen in the great wave of extinctions going on the rainforests around the world. The creatures to suffer worst in this desert region are probably the fish. The endangered desert pupfish is an example of what has been happening to the freshwater fish of the West. In addition, the desert pupfish in the delta is on a par with the Yuma clapper rail in terms of its significance as an indicator of the delta's health.

Once there were five endemic freshwater fishes in the Mexican portion of the delta, according to José Campoy, biosphere director and himself a freshwater fish biologist. Four of these have already disappeared, he told me: the bonytail (*Gila elegans*), woundfin (*Plagopterus argentissimus*), the salmon of the Colorado, or pikeminnow (*Ptychocheilus lucius*), and the razorback sucker (*Xyrauchen texanus*).

The desert pupfish is the only one to survive.

Before the area was settled and the Colorado River dammed, the area around the geothermal plant was a lake of hot and active waters, called Volcano Lake. Overflow waters from the Colorado frequently fed this lake. Mostly, though, it was fed by geothermal springs from this highly active geological area. The springs underlie the huge ponds that were created by the Mexican government when they built an electricity-generating plant over the hot springs for the whole of Mexicali. The area is a surreal industrial landscape of towers and pipes, sheet metal buildings and pumping stations. The location is a patchwork of rectangular lagoons. Some of them are very beautiful, a whole palette of pastel and tropical blues. Some of the empty lagoons are coated with thick layers of salt.

As I stand in the sun, watching Jaqueline try to catch a pupfish in her blue net, I remember reading about a cowboy who visited this area of hot springs in the late nineteenth century. Mud volcanoes were also here. He felt the heat rising up to meet the hot air of the day. He looked down into the spring. "I don't believe," he said emphatically, "that hole is more than forty feet from hades." (Strand 1981, 60)

It is astonishing that fish are swimming in this "entrance to hell." I bend down for a close look at the desert pupfish. It looks like a scrum of tiny fishes. They hang near the rock wall of the pond. They are beige and brown. The males have dark stripes and thick bellies. They are identifiable as desert pupfish, but to an uneducated eye, they might look simply like about 150 chubby little minnows.

But this nondescript and easy-to-overlook minnow is one of the most remarkable fish in the world. We are inclined to think that an endangered species must be fragile. In a sense, it is—as a species. These pupfish have spent at least the last seventy years on the brink of extinction. Still, they are anything but fragile.

The desert pupfish survives only because it can endure the most extreme conditions of any known fish. The water is an example. Jaqueline quits fishing and measures the water temperature. It's not boiling, but it is hot: 40 degrees Celsius, nearly 100 degrees Fahrenheit. Not many fish survive in these temperatures. The desert pupfish holds the record of all fish in the world for the highest water temperature tolerated: 44.6 degrees Celsius, about 120 degrees Fahrenheit. Desert pupfish also hold the record for the lowest amount of oxygen tolerated. Their ability to tolerate salts, says Jaqueline, "is no less remarkable." The fish will also dive into mud that contains no oxygen and is high in hydrogen sulfide as a way of reducing their metabolic rate and controlling their body temperatures.

Jaqueline explains that the Spanish for the pupfish is *pez chachorita del desierto*, a literal translation of desert pupfish. Jaqueline speaks English perfectly well, but she prefers to speak Spanish when possible. This is the last field research she needs to do on her dissertation. She says she is looking forward to moving back to Guaymas when she finishes her degree.

"We call them *pupos*," Jaqueline tells me.

"*Super pupos*," I joke. We both laugh. Super fish.

After a futile half hour in the direct sun, we can take no more. We duck into shade behind the pumping station. The water is too deep. The fish swim easily away when we try to catch them. Jaqueline and the engineer confer. They decide to see if they can find a shallower pond with *pupos*.

I will stay here, I say. I want to try again to catch some fish. Leaning over the water, I create a bit of shade so I can see the fish better. Slowly I lower the

*Using traps with bait, Jaqueline García-Hernandez tries to catch a few "pupos" or pupfish in a freshwater spring for her study of toxic chemicals in the delta.*

net and position it behind a group of four fish near the rocks, hanging in the current that flows into the pumps. I try for complete stealth, moving as imperceptibly as possible.

The sun hammers my back. Drops of sweat fall into the water with small splashes. The fish don't move. Nor do I. I become a statue, ignoring the heat. Suddenly, I am reminded of the green-backed heron I watched stalking fish in the Ciénega de Santa Clara. Herons are utterly concentrated in the moment as they fish. It is the great task of life—to be present to the particularities of the moment and the place. I have a sense of the heron's humpbacked concentration as I bend over the pond, fixed on the pupfish below. The imitation of animals in their most focused moments may be as close to pure being as humans are allowed to come. Several times I make a stab at the pupfish. Each time they easily elude me. I don't much care, except that I want to help Jaqueline in her research.

Strange as it seems, the pupfish's ability to endure even waters as hostile as these in the geothermal plant explain why it has survived. It is one of the pupfish's last strongholds. In the 1970s, fish biologists—ichthyologists—estimated that the desert pupfish had lost 95 percent of its original range. It had been reduced to just a few springs on the Mexican delta and near the Salton Sea. Once, the species was widespread and abundant throughout southern Arizona, southeastern California, and the Mexican portion of the delta. Yet like many of the endemic freshwater fish of the lower Colorado River, it has taken a terrible plunge in the age of dams. Despite its amazing abilities to survive any of the extremes the desert could heave at it, the desert pupfish has struggled to survive the dewatering caused by the last century's dams and irrigation practices. In 1986, it was officially declared an endangered species in the United States. The foremost expert on river fishes, W. L. Minckley, declared in the 1970s that the pupfish was on "the brink of extinction."

The surprising growth of the Ciénega de Santa Clara offered unexpected hope for the pupfish. In fact, the Ciénega may sustain the largest population in the world of the endangered pupfish. Like the Yuma clapper rail, the pupfish now has better prospects for survival in the Mexican delta than it does in the United States—the delta has become the last best place for this once-flourishing species.

In fact, there are several species of *Cyprinodon*, or pupfish, that occur in the American Southwest and northwestern Mexico. For example, there is a species closely related to the desert pupfish occurring in Quitobaquito Springs and the Sonoyta River in southern Arizona and northern Sonora, Mexico. There are several other species that occur in California, including Death Valley. They all represent surviving populations that have evolved into their own species through geological time. Once they all were part of the same population of fishes, when a Miocene–Pliocene sea covered the area.

You can think of these small fish as relics of prehistory themselves. You can even think of them as surviving in fossil water—water laid down in the aquifer during geological time. You can think of them, in a way, as fossil fish, except that they are also marvelous examples of the way that nature evolves, adapts, and survives. More than anything, the pupfish are marvels. They are living oxymorons: desert fish.

They are also examples of the way western culture has destroyed life in the desert. Desert fishes have suffered horribly in the western United States. The fish of North America are the most disturbed and destroyed fish in the world. Fully 70 percent of all the endangered fish in the world occur in North America—and over half of the 254 species of fish on the endangered or threatened list in North America occur in the West. This, even though the West is much drier than the eastern part of the continent and has many fewer fish—about 150 freshwater species versus 600 in the East. And the worst place for these endangered fish on the entire continent is the Southwest—the lower Colorado River.

Desert fish evolved in isolation. They live in islands of water surrounded by seas of sand. The Colorado River, for example, has the largest number of endemic species—species that evolved here and nowhere else in the world—than any other North American river. The most vulnerable of all species, as the scientific field of island biogeography has demonstrated, are endemic island species. The arid West with its freshwater fish is on a par with the Hawaiian Islands and the state of Florida for having the highest rate of extinction and endangerment in the country. Of the 68 species of birds unique to Hawaii, 41 are already extinct or on the verge.

Of the 150 or so full species of freshwater fish described in the American West, 122 are endangered, threatened, or in trouble. As W. L. Minckley wrote in 1991:

> Worldwide clearing of rain forests has had a parallel in the destruction of natural freshwater habitats in the American West for decades. Despite the emergence of an enlightened public, determined efforts by agency and academic biologists, favorable legislation, and other efforts, a wave of native fish extinction is continuing in arid lands. (Minckley 1991, 17)

The pupfish call into question our headlong population growth in the desert, along with the water exploitation that leaves the rivers, especially the Colorado River, sucked dry. The desert pupfish is an endangered species that points the finger directly at water policies along the Colorado River in the United States. These fish are a symbol of a culture's overweening exploitation of the water in the desert.

In the United States the main victims of the new regime on the Colorado River are four species. The pikeminnow (squawfish) reached six feet long and weighed up to 100 pounds. Strangely, it was the largest freshwater minnow known to science. Wild populations are extinct in the lower Colorado River, although it still persists in the Upper Basin. In the Lower Basin of the river, the last pikeminnow was collected in the 1970s in the Grand Canyon.

The bonytail chub is nearing extinction in the Colorado Basin. Reaching about two feet in length and a weight of three pounds, it is a lot like a fast-swimming mackerel. It took thirty years to find even fifty bonytails, the species is so rare. Yet in the 1930s it was one of the most common fish found in the lower Colorado River.

Populations of the razorback sucker are also spiraling toward extinction and are deeply endangered. This species grew to three feet in length, reaching eighteen pounds. Historically, it was so abundant that it was harvested from irrigation canals and used for human and livestock food, and fertilizer. There are approximately ten thousand left in the lower Colorado River basin, with an estimated five hundred persisting in the Upper Basin. The razorback sucker is extinct in Mexico. There are recovery plans in place for all of these fish, plans that are being implemented in varying degrees.

As native freshwater fish have disappeared, a new order has emerged. The dams replaced the rushing waters of the Colorado River with still, clear reservoirs. Into these artificial lakes, new sport fish were introduced. These nonnative fish—"exotic species"—include trout, bass, carp, and catfish. Most of the native fish were not aggressive. The introduced fish are—they are "aquatic pigs." Most devastating, the introduced fish eat the eggs and the larvae of the native fish, which means there are few young native fish to continue the species. The gradual attrition of adult stocks causes the native species to go slowly extinct.

Not only can desert pupfish survive in environmental extremes, they are also very tough. The males are extremely territorial. They will take on anything—other pupfish males, other species of fish. In the Ciénega, for example, bass have been introduced. Little two-inch male pupfish will attack a huge bass that's going after his mate's eggs. The results of these battles are not always, as one biologist put it to me, "favorable to the pupfish."

Despite their best efforts, the pupfish continue to suffer severe stress from loss of habitat and from introduced species. Its survival, like that of desert fishes throughout the Southwest, is very much in jeopardy. The desert pupfish is one more example of why the United States and Mexico should develop a binational approach to delta recovery. So far, the fish's survival is largely a testimony to its own toughness.

The fish's luck extends to selenium, as well. This mineral is found in volcanic soils. Humans, in fact, need it in small amounts. At higher levels, it becomes toxic. The Colorado River always has high levels of selenium. There are high levels of selenium, boron, and arsenic in birds in some wetlands in the delta, most notably the highly polluted Río Hardy wetlands, which will be discussed more fully in a later chapter. Dissolved selenium in these wetlands is up to fourteen times higher than the Environmental Protection Agency's criterion for protection of freshwater fish (Alvarez-Borrego 1999, 1).

In the Kesterton Marsh in California there was a large die-off of birds from selenium. Jaqueline's research was aimed at determining levels of selenium to help in plans for the restoration of wetlands in the delta. Despite the high levels of toxins in parts of the delta that need to be addressed, Jaqueline's tests would determine that pupfish do not have elevated levels of selenium.

For an hour at the geothermal plant I bake in the direct summer sun while I try to catch these little pupfish for Jaqueline's research. She needs fifty fish for her research. Occasionally I rest in the pump station shade. The *pupos* prove elusive. I finally catch one, a sick one, drifting slowly by. Then I snag another. Two. Still, I am happy to catch even these two.

When Jaqueline returns, I offer her the two little fish. I call myself "Martín Pescador," the Mexican name for the bird we call the kingfisher.

She decides to leave a trap in the pond with a bit of food in it. Tomorrow we will return and find the trap filled with little fish. She will choose several and put the rest back in the artificial blue lagoons.

We leave the sweltering geothermal plant in the late afternoon. The lives of these tough little fish are compelling. They have survived the onslaught of western culture that has taken over their waters, both river and springs. While many other fish have gone extinct, they have demonstrated remarkable tenacity. Now they hold out in artificial refuges like the Ciénega and these geothermal lagoons. These amazing little creatures are genuine super fish.

FIVE

# A Dream of Clams

ENDANGERED SPECIES: Colorado delta clam, *Mulinia coloradoensis*
PLACE: Shelly Islands and Gulf of California mudflats (west side)

WITH A SOFT and yielding thump, the keel of the boat smacks mud in the shoaling tide. Karl Flessa leaps into water up to his knees. The rest of us follow. We each grab a couple of bright blue wet bags and lug our gear through the water to a small island of seashells.

The island is blindingly white. It rises only a few yards above the tidal mudflats, which stretch for miles in every direction. Called a chenier, the island is only about the size of a football field. It is made up of millions of wave-tossed and wind-polished white shells in a loose pile. No dirt. No mud. No solid footing. We scramble up the steep side of the *isla*. The shells give way like sand, only clattering. Once on the top, we pitch tents on the shells. The fishing boat turns and leaves, cutting a wake through the gleaming sea.

And we are all alone.

You will not find this tiny island on any official map of the Gulf of California. It is too ephemeral, too unknown, too unrecognized. In a sense, it is not quite here, an almost invisible part of the landscape. You can see it, though, on satellite photographs of the Upper Gulf, a thin white line several miles out from the shore, alone in the mudflats on the west side of the delta. That's how Karl showed me where, exactly, we were going. Local fishermen know about these islands as well. This one has a local name. Ramon Soberanes, the fisherman who ferried us here, calls it *Isla Sacatosa*. It is one of several small shelly islands and beaches that have formed on the delta since the dams went in on the Colorado River.

◄ *The vast mudflats of the northern Sea of Cortés, on the western side of the delta, once supported an astonishing abundance of clams and sea life. The mud was thick and rich with nutrients, but dams have reduced the diversity of life by 95 percent.*

▲ *The beautifully shaped shells of clams suggest both humility and home—and desire.*

93

Since discovering these islands on a fly-over with Sandy Lanham, Karl has been coming to them for several years, camping for a few days at a time. They are one of the field sites for his study of clams in the Upper Gulf. He is making these otherwise unknown islands into landmarks in the new and modern geography of the delta. A geologist at the University of Arizona, Karl has fair Germanic features and a trenchant sense of humor. He has long been interested in the way the fossil record does and does not represent the "facts" of the earth's history and evolution. He knows rock 'n' roll very well and is especially attracted to the Talking Heads, whose lyrics he will sometimes quote on the philosophy of science. "Facts just push the truth around," he wrote in one article, quoting lead singer David Byrne. Karl describes himself, geologically speaking, as a "soft rocker"—interested in sediments. The geologists interested in mountain formation and tectonic plates? "Hard rockers," Karl smiles.

Karl had been studying the fossil record of the delta when he found himself attracted to what the huge clam beds in the delta might be able to teach about the biography of the Colorado River. Karl turned highly sophisticated research techniques to the study of clams. He is delightfully aware of the ironies of his research subject: the big scientific education and sophisticated scientific equipment trained on an anonymous little clam virtually no one has ever heard of before.

"Clams are not charismatic," he laughs. "But they are funny."

They are also an indicator of the health of the delta.

You might say that he has taught the clams in this area to speak. The difficulty for ecologists in evaluating the modern condition of the delta has been the lack of biological information from the period before the dams were constructed. The place just had not been well studied before the dams, except in ways that helped farmers and engineers. Ideally, scientists want information from both before and after periods when they are trying to study possible changes caused by a significant event. That's how you can make clear assessments of the environmental consequences. Though scientists from various fields, as we have seen, have been studying the delta intensively in the past fifteen or so years, no one had documented the biodiversity of the delta before the dams.

That's what Karl is doing. In the process, he is providing scientific proof of the devastating ecological impacts wreaked by the upriver dams on the delta ecosystems. The shells of the clams that have formed these islands hold the key to understanding and documenting that before the dams, the mudflats here were once fabulously productive. The beautiful islands in the Gulf on which we are camped are the tangible remains, the evidence, of a huge wave of extinction that has taken place here. It is a clam graveyard.

With tongue firmly in cheek, Karl parodies the categories of geological dating when he describes the two major relevant periods of time he studies:

"Pre-Damozoic" and "Damozoic." Karl's clever time periods identify with humor what is increasingly recognized as the central truth about the delta. Time in the delta is divided really into the "pre-dam" age, and the period we live in, the "dam age."

His humor can be biting, too. He calls one paper about the loss of biodiversity here, "The Silence of the Clams."

Combining chemistry with geology and paleobiology, Karl has found that one species of endemic clam in the Upper Gulf speaks for an entire ecosystem. His research is the best evidence yet on the biological fertility of the pre-dam delta. The lesson of the clams is that the delta was once, as Karl puts it, "teeming with life—fish, shrimp, clams, oysters."

One particular species of clam, *Mulinia coloradoensis*, has been the object of Karl's research. It hardly seems a creature likely to take on the entire social project of the Bureau of Reclamation and the creation of the richest civilization in the history of the world in the American West. The little *Mulinia* clam evolved in the Upper Gulf. Like many other marine species in this delta ecosystem, it grew to be dependent on the unique combination of salt water from the sea and influxes of freshwater at key times of the year, during the creatures' growth and development.

There are other species of *Mulinia* clams. But *Mulinia coloradoensis* is endemic to the Upper Gulf—the only place in the world where this clam can be found. Karl has determined through study of the clamshell characteristics—the traditional way of distinguishing clam species—that *Mulinia coloradoensis* is a unique species. The research also suggests that this clam is genetically distinct.

We are here on Isla Sacatosa so that Karl can collect more of the *Mulinia* shells, along with other species that he can use for comparative studies. The hard shells do not disappear like the bones of other creatures. Using the clams' shells, he has been able to generate specific numbers on the astonishing productivity these river-silt mudflats once achieved. And just as important, he has been able to prove that the former abundance was created by the flows of the pre-dam river.

He explains as we gather around the camp stove to make hot chocolate.

We are sitting in the middle of the most extensive tide flats created by the river, he explains. The tidal action and the currents of the Gulf carried most of the silt in the river in a westward flow. The very fine silt was deposited here over the millennia into a sheet of extraordinarily broad, rich mudflats. The mudflats from the river extend 65 kilometers south of the river's mouth—about 40 miles. They are miles wide, as well. Robert Thompson's study of sediments demonstrated the enormous size of these west-side mudflats. They consist of 2,000 square kilometers, or 722 square miles, of mud that goes down 16 meters (Thompson 1968, 1).

This whole area is washed by the some of the largest tides in the world, reaching thirty feet. The sea laps in and covers the entire area. And the rich soils and seawaters made these clam beds teem with life—clams, snails, worms. "Our official estimates indicate that there are the remains here of *two trillion Mulinia* clams," Karl says. "It was just thick with them here."

"That's a lotta clams," he says. That's the number that have died and constitute the shelly islands, or all the clams that have lived in the area in the past one thousand years. Karl is sure the estimate is too low. He actually believes the number is closer *five trillion* clams in these mudflats all around us.

Using radiocarbon dating and amino acid testing, he is able to focus on the clams that lived here before the dams, from 50 to 950 years ago. He studies the shells of dead clams. That's why he calls his research center *Centro de Estudios de Almejas Muertas*, the "Center for the Study of Dead Clams." Since he knows that clams live about three years each, he knows that there are 333 generations of clams he is studying. Karl calculates that at any one time there were about *six billion Mulinia* clams alive on the delta mudflats.

How did he generate the original numbers?

"I had an undergraduate count them," he quips. Then he says, seriously, that he and his researchers simply did the calculations. They have been digging clams here for some time and they have been able to determine the density of *Mulinia* clams in the mud. There were about fifty clams per square meter before the dams went in.

A different species of clam has moved into the mudflats now that the dams are in. It is called *Chione cortezi*. All the live clams now average only three clams per square meter.

"We estimate the marine systems of the delta have lost about 95 percent of their former richness and diversity," Karl says.

I ask Karl what has happened to the *Mulinia* clam in the delta. Has he found any of them still living, or have they gone extinct? He tells me they have found about twenty living *Mulinia* clams at the south end of the muddy Isla Montague. That's it—from six billion clams alive in this region as recently as, say, one hundred years ago, to twenty.

He hastens to add that this number indicates only the number he has found. He is sure there are many more living, on the order of twenty thousand. That's a loss of as much as 99 percent of the population. Based on his research, Karl is proposing the endemic delta clam, *Mulinia coloradoensis*, for official listing as an endangered species.

The even more interesting story is how Karl has been able to prove that the dams were the cause of the decline. That's what he would show me the next day on a hike that we later came to jokingly call "the delta death march."

This afternoon I let myself enjoy the strange and lonely beauty of the island we are camping on. The shelly islands are the visible signs of both the erosion of the delta and the death of its former abundance. As we have seen, the delta is eroding. It is literally in retreat because the river is sediment starved. The waves and currents of the sea are washing the delta away. But they are only strong enough to carry away the fine silt and mud. The shells get left behind. With each rising tide, a few of the shells get moved down the beaches. Slowly these exposed shells grow into islands.

But the landscape is still beautiful in the soft light of the afternoon, which hangs like a sheer blue veil over the wet mud and sea. In the distance, the mountains of the Baja California peninsula define the far horizon. They were created out of long-vanished volcanoes, partly eroded and smooshed up out of the collision of tectonic plates. Their sloughed dirt combines with the silt from the river to form the abrupt flatlands at their bases, called *bajadas*. The mudflats around the island are full of life and activity. The mud is full of holes created by crabs and clams and worms. Seabirds and shorebirds of all kinds are working the mud. I watch a huge long-billed curlew, for example. This is the largest shorebird in North America and has a very long downwardly curved bill—perfect for probing deep mud. I watch one curlew stab repeatedly in a single spot, then pull out a crab in its pincered beak. It gulps several times to choke the crab down.

I am standing on an island that is the proof of a massive die-off, yet there seems to be so much life all around us in the mud. I can't really imagine what this place must have looked like a hundred years ago. And I realize in that instant that what we have done to wildlife has made it impossible for us to genuinely appreciate the true meaning of abundance. We have reduced the animals to a fraction of their former levels. And then we have come to believe that this reduced level is the real level, that these diminished creatures are the real creatures.

Among the saddest victims of the wave of endangerment and extinction, these clams suddenly teach me, are our own imaginations.

By noon the next day we are in the middle of nowhere, waist deep in mud, flinging clams. Using GPS coordinates, Karl has led us to a precise spot in the mudflats, five miles from Isla Sacatosa. It has been a long hike to get here, hauling gear. We are in the midst of the delta's guts. The mudflats are baked out and cracked in geometric shapes. They ripple in hallucinatory patterns, monotonous and desolate, for miles all around. No geological relief. No escape from the sun. It is an empty world of excesses—too much sun, too much mud.

Except that you realize as you dig that, if this place is barren, it is a fecund barrenness. At least, it was.

I have dug a hole in the mudflats. The tide has been receding all morning. The ground wavers in the hazy heat. Off to the north I sometimes imagine I can see the mouth of the Colorado, but it is only a mirage. Just beneath the surface,

the clamshells are thick and dense, packed into the cool and still-moist mud. They belie the seeming desolation of the delta. The *Mulinia* shells are delicate and shiny white, smoothly enameled and spotted with brown. Karl wants us to collect twenty-five *Chione* shells, which are much less common. Karl also wants us to dig down so we can find shells that have been lying in constant temperatures. The *Chione* are like ridged potato chips, large and rough. Karl says the *Chione* is "well-behaved for our purposes."

I look back at Karl, in his own hole. All around him is nothing. And he is flinging clams and mud through the air.

His purpose is to test the clams chemically, back in his lab in Tucson. Karl also has three students with him on the field trip. Finishing his Ph.D. dissertation is David Goodwin, a veteran of many excursions. Bernd Schöne is a postdoctoral student from Germany and has been here once before. Kirsten Rowell is about to enter Karl's lab.

Karl had decided upon this location in earlier years. Before he had begun camping on the islands, he would drive out into the mudflats and then hike from the car. "After I got stuck in the mud one time really badly," he says, "I learned to be a lot less gringo about things out here." He also learned to take a boat to the islands and camp within hiking distance.

As we hiked the mudflats, we encountered other signs of life. We found a pair of burrowing owls on one small ridge, miles out into the flats. We found coyote tracks crossing a particularly barren and cracked stretch of mud. We found bones of dolphins, lonely and disintegrating in the sun. And we found glass fishing balls, some exquisitely handblown. It is strange to find beautiful reminders of human handicraft, even in such a place.

Through the tests Karl conducts on the shells we will gather, he can detect what he calls "the signature of the river." Through that, he can tell what the role of the river's freshwater is in producing the clams in these mudflats. The shells of these clams are calcium carbonate, $CaCO_3$. Clams manufacture their own shells, taking oxygen out of the water and secreting it into their growing shells. Like trees, clams lay down their growth in rings, or ridges. Oxygen comes in varying isotopes. That is to say, the number of neutrons in the oxygen atom varies. And the oxygen isotopes in the salty Gulf of California differ, very slightly but also measurably, from the oxygen isotopes of the freshwater in the Colorado River. Through isotope analysis back in the lab, Karl and his students can identify the oxygen isotope from the river. It is a "tracer" of the river, as Karl puts it.

We are likely to think of clams as anonymous creatures, alive but not really having lives. Yet Karl's research is so sophisticated that it can detect even the most minute growth phases of these clams. They are not merely symbols of the fertility of the former river. They are creatures with lives in a very complex relationship to the river and the tides of the sea. They have their own marvels to tell.

"In dead clams," Karl says, "we can figure out how old they got, how fast they grew, what time of year they spawn, and the conditions under which they grew the fastest. It's pretty nifty, huh?"

The German postdoc, Bernd, is often silent. He loves building huge fires on the shelly island. And he frequently patrols the beaches for clams. He is looking for research subjects. He is doing the research that enables the construction of what you might call the "biography of a clam." According to Bernd, "Each ring on the clam, even microscopic rings, can be deciphered." Clams grow in summer, he explains, with the high water from the river. They stop growing in winter, when there is not as much river water. They even have growth spurts when water comes in. By cutting the shells, polishing them, and putting them under a very powerful microscope, Bernd can study the growth of the clams at extremely subtle levels.

"They have various growth cycles," he says. "Clams stop growing every two weeks on the lunar/tidal cycle. Then they spawn. They also have growth cycles each day. We can see growth patterns of the daily tidal-cycle—one mini-ring per high tide."

By correlating growth cycles and oxygen isotopes, Karl and his research students and colleagues know that the *Mulinia* clam depended on the freshwater of the Colorado River for its survival and its growth.

"Some day we'll be able to tell what these clams had for breakfast," Bernd jokes.

Karl's research on these clams has not only demonstrated the impoverished biological condition of the intertidal zones of the delta. It has the promise of being applied to other endangered marine species in the delta. Crucial riddles could be solved.

Kirsten Rowell, for example, will enter Karl's lab to study the totoaba, a huge fish of the Upper Gulf. Once it was abundant. Now it is deeply endangered, no longer part of the fishery. The totoaba has an otolith, a unique bony structure in the inner ear of the fish. It's often called an "ear stone" and it, too, is calcareous. It has aragonite, which the fish uses to record sea temperatures and the amount of freshwater in its habitat. Karl describes the otolith as a kind of environmental "flight recorder." Otoliths bear isotope traces as well, like clamshells. By a careful study of the totoaba otolith, Kirsten hopes to be able to determine if part of the reason for the decline of the species can be attributed to the loss of freshwater flows into the Gulf from the river. Perhaps these studies will prove that at a crucial stage of the totoaba's life, it required freshwater from the river, like many other species.

Deeply endangered species are notoriously difficult to study. One powerful case in point, from the delta, is the vaquita, the "little cow." The vaquita is the world's smallest marine mammal. It is a small porpoise found only in the Upper Gulf of California. It is also the rarest marine mammal in the world. Only documented by Kenneth Norris in the late 1950s, the vaquita is also deeply endangered. There may be as few as two hundred of these creatures left, rarely seen except when they turn up dead in fishermen's nets. Part of the reason for establishing the biosphere reserve was to protect the vaquita.

Very little is known about this intriguing creature. It is too rare to study. No one has been able to determine if, for example, like other species in the Upper Gulf the vaquita required flows of freshwater at a crucial stage in its development. Karl plans to study the vaquita for oxygen isotopes as well, to try to determine the role of freshwater in its life cycle. Since abundant skeletons of the vaquita exist in museums, he has access to the dolphin's teeth—which also carry the signature of oxygen isotopes.

Karl has just received major funding from the National Science Foundation to study *Mulinia*, totoaba, and vaquita for isotopes. "We think we can determine how much river flow will be required to restore these three species," he relates. Through his geological and biological work, Karl is developing an exciting new field of conservation biology with the potential to help us understand the natural history of deeply endangered creatures. He calls it "Conservation Paleobiology." This unassuming clam has led the way.

Karl says this is his first brush with research that has policy implications. "I avoid comments that lead to policy about water," he says. "But I like it that others will use this research and its implications. The delta has died for lack of freshwater."

The obvious implication is that water is key to the productivity of the delta—here in the mudflats as in the marshes and wetlands. As we will see later, it is key to the productivity of the marine system as well. Local fishermen have long been claiming that the dams destroyed the productivity of the Upper Gulf. Karl's research definitively establishes a cause-and-effect link. His findings corroborate another important study, which demonstrated the same link: shrimp catches are higher in the years following controlled releases of river water from the dams.

These vast intertidal mudflats were once clam beds of a size and complexity that at once inspire and humble the imagination. Perhaps the best way to get a sense of the scale of things in the delta is to walk through it. Combined with the morning hike and the digging through clamshells, the return hike is grueling. The tide is coming in. The mud is wetter. We sink deeper with each step. It is only five miles, but a slog through the delta mud is an unrelieved endurance test for all of us. Exhausted, we slough back into camp in the early evening. We good-naturedly call Karl a tough taskmaster, likening this scientific excursion to an old "desert death march" across the delta, a *jornada del muerto*.

After a quick dinner, I climb into the tent. The rigors of the delta have impressed themselves onto my weary body. Mindless as mud, I crawl into my sleeping bag. As I lie there, I think about one of the great *jornadas* through exactly this part of the delta. It was by another of the great figures in the history of the delta, and another of the few white men to love the place, Godfrey Sykes.

And then I collapse into my bed of clams.

Before the recent spate of delta studies—like Karl's—you could count the number of white men or women who loved the delta for its own sake on your two hands. Very few North Americans or Mexicans have been able to see the natural values of the delta—to see, as it were, the richness of life buried beneath the surface of the mud. Aldo Leopold was one, writing beautifully of the green belt of the river through the delta in his "Green Lagoons" essay. John Van Dyke wrote more eloquently than anyone of the delta's deserts; he was one of the few who could see the beauty of this daunting and even terrifying land.

However, no one of European descent, I suspect, loved the delta more and spent more time here than Godfrey Sykes.

Sykes did not develop a reputation as a great thinker about environmental matters, as Leopold did. Nor was he known for his aesthetic sensibility and the beauty of his prose, as was Van Dyke. But Sykes was profoundly dedicated to the delta and was the greatest of its students before the modern round of research. Sykes discovered the delta almost by accident, wandering around the world and

*The vast tracts of mud make for weary hiking, as the great student of the delta Godfrey Sykes discovered on his nearly disastrous* jornada del muerto *through this same region over 100 years earlier.*

through the American West. From almost the first time he saw it, the delta became the central theme of his life. For forty-five years, between 1890 and 1935, Sykes studied and mapped and wrote about the delta. He was its first great expert, the delta's greatest authority, the region's most intimate student.

For modern researchers in the delta, his book, *The Colorado Delta*, is always invoked with the deepest respect. Published in 1935—the same year that Hoover Dam came on line and that marked the real death of the pristine delta—*The Colorado Delta* is a careful study of the geography, hydrology, and sedimentation of the delta, written by a man who was himself an engineer. The book has a formality and distance to it that are the result of a scientist's and pioneer's sensibility. *The Colorado Delta* offers many references to the abundant plant life in the delta, with impressive photographs of forests and plants. If the book lacks anything, it is that it says next to nothing about animals and bird life in the delta.

Yet it is also, and more movingly, a kind of elegy to the delta's demise—everywhere *The Colorado Delta* is suffused with a soft regret and sad farewell for the "terra incognita" and "untraveled wilderness" that Sykes originally found here. It is a scientific book written out of love. As an engineer at the time, Sykes was unusual, if not unique, in viewing the river not only as a problem to solve but as a unique and inspiring ecosystem.

This view of the delta comes across even more clearly and touchingly in another book written by Sykes. My own view is that *A Westerly Trend*, published in 1945, is a much more engaging account of his relationship with the delta and deserves to be better known for its description of this formative time in the Southwest's history. The book is autobiographical and so conveys more of Sykes's feel for the river and the delta. Born into a small English village, he slowly worked his way west, lured by the image of the American cowboy. When he discovered the Colorado and its delta, he found his passion—moving his family to Yuma at a time when most viewed the town as a forsaken outpost on a nearly uninhabitable frontier. He even took his family on vacations down the Colorado into its delta.

Nobody traveled the delta more than Sykes—except for the indigeneous people who had lived there for generations. He describes the river as having an "unrestful and changeable spirit in that particular area"—"a perverse, wayward, and temperamental river." The following paragraph beautifully summarizes his relationship to the delta, taken from a description of one of his vacation camping trips with his family. It is lovely and loving account in a way few written descriptions of the delta are:

> We camped at times for a week or more at the same place and at such semi-permanent homes I generally built a ramada of willow poles and branches for my wife and children to rest and play under. These stationary periods also afforded me opportunities for indulging what was becoming a major obsession: investigating and poking into the many peculiarities of the deposition of silt, the formation and erosion of bars, shoals, and banks, along this most interesting and extraordinary of rivers. In the course of our wanderings and restings, opportunities occurred for carrying these stray and casual notes down through the delta to the tidewater during the Winter seasons terminating in 1903 and 1904. This region was still almost unknown and seldom entered save by the indigenous Cocopah Indians and a few Mexican cow-folk, few of whom were greatly interested either in Geography or Hydrology, and my notes, although fugitive and incomplete, soon became rather interesting as matters of geographical record, when the Colorado River, following upon many years of comparative quiescence and stability in alignment, suddenly decided in the Autumn of 1905 to abandon the terminal hundred or so miles of its course and head off through the wilderness to the northwestward . . . . (Sykes 1945, 245)

The final portion of this description refers to the great events that changed the course of the delta forever—the diversion of the Colorado into the great desert that is now the Imperial Valley, and the subsequent engineering disaster in which the uncontrollable river created the Salton Sea.

Sykes knows that the delta is doomed by the engineering plans upriver, that he is writing about the "end of a great delta" (Sykes 1938, 241). His work is essentially an elegy for a delta that will be no more, "with only a man-regulated and restricted amount of water for future pranks in the ghost of its delta" (Sykes 1945, 310).

*The same mudflats traversed by Sykes. Here Karl Flessa leads his team to a remote spot to collect clamshells for study.*

When Sykes was exploring the delta, very few whites, as he says, visited the area. Even John Wesley Powell, the hero who first conquered the Grand Canyon and put the Colorado River on the American map, never visited the river's delta. Most of the interest in the delta in Sykes's time centered on finding ways to exploit it, using the river water to convert it to farmlands.

This was not the kind of landscape that most people knew how to love and appreciate—unless it was converted to moneymaking projects. Sykes was different though. Even when he encounters conditions in the delta at their most extreme—floods, blistering heat, near starvation—he always describes it as a beloved place. One of his most amazing experiences in the delta—where he faced the delta's worst rigors—took place exactly in the vast and dry mudflats where Karl and I were camped.

For Sykes's friends, his ordeal was nothing short of heroic. His botanist friend, Daniel T. MacDougal, wrote an article that he called "The Delta of the Colorado River." His writing provides an example of how most people viewed the delta—as a landscape of hardship and disaster. As MacDougal writes in his essay, "Perhaps not more than one in three . . . adventurers have escaped hardship or disaster in some form" in visiting the delta. MacDougal says that Godfrey Sykes himself endured a *jornada del muerto*, or death march, that had not been "attempted before or since by white or Indian, which was won out by an intimate knowledge of the ways of the desert and by a sheer capacity for endurance" (MacDougal 1906, 3). Sykes tells the story in a very different way in *A Westerly Trend* and it's a wonderful narrative. That he survives is astonishing testimony to his knowledge of the area and his ingenuity.

In 1890, Godfrey Sykes and a friend built their own boat and sailed down the Colorado River, through the delta, out the mouth, and south along the western shore of the Gulf of California. He describes the fascination with adventure that drew him onward:

> The river below Yuma, the great tidal estuary, the vast expanse of mudflats about its mouth, and in fact the entire upper end of the gulf, had reverted to the condition of a wholly forgotten region . . . at the time of our exploratory expedition; but this, of course, was an added attraction and we gaily headed down into the unknown in search of such adventures as we might encounter. . . . (Sykes 1945, 214)

No good maps were available to Sykes. He was using a wholly inaccurate map dating to 1826.

They sailed more than 150 miles down the Gulf. One evening while enjoying a fire on the shore, a gust of heavy wind blew cinders into their boat. It caught fire and burned the whole front end away. They were able to save only the water jug.

Sykes says he knew they were in trouble. They had seen only one small group of natives since leaving Yuma. The only water hole they knew of lay over one hundred miles north of them, near San Felipe Point. They had no choice. They began the long trek over the beaches heading north. For four days they hiked, covering about twenty-five miles per day over "the soft ground near to the coast."

They found the water hole. A dead coyote was floating in it. They "bailed it out" and with a good stiff upper lip "took some good long drinks of warm and rather unpalatable, but refreshingly wet water" (Sykes 1945, 222).

In the region about San Felipe they also found a big bed of luscious oysters. For three days, they stayed near the oysters and the water, "to give our fallen arches a chance to rest." The last night here, Sykes's partner managed to shoot a coyote. It could "only be called edible by straining several points." They tried to eat "coyote cutlets" for breakfast—but the flavor of decomposed fish in the coyote's diet was insufferable (Sykes, 1945, 222).

*The millions of dead Mulinia coloradoensis on Isla Sacatosa create a small island in the gulf.*

The oysters sustained them on the long seventy-mile hike through the vast sweep of intertidal mudflats between San Felipe and the mouth of the Colorado River. Sykes calls it "soft and troublesome walking," an understatement of magnificent proportions. They survived and followed the river up to Yuma. There, "rather ingloriously, our projected voyage to far away places came to an end" (Sykes 1945, 224).

The adventure goes a long way to defining the early romance of the delta as a scene of heroic striving. Sykes tells a wonderful story. A sense of the strangeness and otherness of the delta shows in his narrative—it is "perverse," a "semi-navigable desert." Yet it becomes more than that in Sykes's story. He never complains in telling the story, never bemoans his condition. His delight in the desert delta and its adventures shine in his handling of the story. Nor is Sykes merely passing through the delta, never to return. He comes back over and over again. As he does so, he moves to a deeply affectionate relationship with this place. At the end of *A Westerly Trend*, he shows his feelings explicitly:

I confess that I have much the same sympathy for my old friend,
the sometimes wayward, but always interesting, and still unconquered
and untrammeled river of the last and preceding centuries, that I have
for a bird in a cage, or an animal in a zoo. . . .(Sykes 1945, 313)

He bore witness as engineers "hog-tied" the river delta. In *The Colorado Delta*, his language is more distanced, more like the engineer that he was. One detail of his concluding remarks about the delta is particularly revealing, though. He writes that the new "clarified and controlled stream" represents "the new order of hydrological conditions along the lower course of the river." And he marks the date of the delta's death: ". . . for all practical purposes the closure of the construction by-passes at the Boulder Dam [later renamed Hoover Dam], on February 1, 1935, has spelled the end of the growth of the delta body as a major phenomenon" (Sykes 1937, 174).

Sykes is speaking geologically here—the closing of the bypasses caused Lake Mead to form, trapping the river's sediment. From a geological perspective, this is the death of the delta. Yet water could still come to the delta once the reservoir filled. It is not the dams, per se, that destroy the delta. The delta can still receive water, just without sediments. What the dams do, though, is control the water as well as trap the sediments. They regulate the flow of the river. They make it possible to divert the water into canals and aqueducts, the drinking straws in the river. And these prevent water from reaching the delta.

For Sykes, the entire future of the delta is buried in the closing of the dam's bypasses. In his engineer's and geologist's precision there is a genuine poignancy. He wrote what was tantamount to the delta's obituary.

〽

After two nights on Isla Sacatosa, we have moved to another island of shells. This one is actually a cluster of very small and very low cheniers. We have more work to do here in the surrounding mudflats.

But we have free time as well. I find myself compelled by a task that is not scientific. I want to see how it is that you can come to care about an apparently hostile place that has also been so badly abused. I am not sure that we try to save landscapes that we think are destroyed. We save landscapes that we can invest with dream and desire. I want to see if I can find any of that on these fragile *islas* of shells.

What I am trying to do is to find an object or the image that will open the door of this landscape to me. Early one morning, I hike along low edges of the chenier. The sunrise was stunning in its deep reds across the tideflats. But under the cover of gray winter clouds, I am taken by the clamshells on the cheniers.

I realize that there are several species, many of them of stunning beauty. They have a soft radiance in the early light, a shining evanescence, like the islands themselves. The most exquisite of the shells are coated on the inside with a kind of enamel. Each shell is unique in the shade of its enamel, but they all shine with a pastel, mother-of-pearl opalescence. Some are gray, others cream, and the most lovely are a delicate copper. Called jingle shells (genus *Anomia*), local Mexicans string them together as delicate wind chimes. Cockleshells also lie on the chenier. They bulge with a perfectly cupped roundness, delicate and precise. Scallops are scattered across the beach as well. Each of these shells seems almost iconic, platonic representations of the shell. They come in soft pastel blues and deep, reddish hues that verge on purple. Most beautiful in their sensuous curves are the gastropods, the shells that spiral outward in a sweep of lines.

The shells are mathematically precise, the perfect coincidence of form and function in an ideal shape. Their shapes are closely related to Fibonacci numbers, in which spirals grow outward in logarithmic patterns. Each of the shells is an example of a kind of animal geometry. Many creatures grow like this—clams, snails, fixed and round-bodied creatures. Suddenly I find myself immersed not in a world of desolation, but a world of pure geometric forms. They seem among the most exquisite forms, wonderful to the mind, beautiful to the eye.

On my knees, I simply touch the shells, run my fingers over their smooth insides, feel the roundness of their inner surfaces. The entire delta is reduced to these polished shapes, the smoothness and rough ridges of their textures. It may be that the grand and the spectacular draws us most quickly in nature. But I often find that the best way to move beyond simply looking at a landscape, to begin to see it more fully, is through investigating its small parts. When each thing is known and loved because it is small and humble, the world opens up to us.

I am reminded of one of the great images of shells in western art. In the fifteenth century, Botticelli painted Venus, goddess of love, rising from the sea with a delicacy of color and clarity of line that make the image irresistible. Her hair blows in soft curls in a sea wind. She rides on stylized waves in a huge cockleshell. Interestingly, one of the first clams named by Karl Linnaeus was given the genus name of *Venus*. It was not the cockle at the goddess's feet. But the name has been adapted to a diverse group of bivalves, called "venerids," that includes the genus *Chione*, the clams we collected in the delta.

Shells have long been associated with the intimate beauty of the earth. In Botticelli's *Birth of Venus*, they are connected with origins, emergence from the sea, even the fertility of spring. They are containers of an eroticized and seductive beauty, a beauty worthy of the goddess of desire. In their rounded shapes and lovely feel, they have been associated with the feminine and connect us to beautiful sea changes. There is even the suggestion, in the painting, that Venus herself, love herself, was born from this ocean's cockleshell. Perhaps we can learn from shells a new way of imagining ourselves in the world—not conquering the world, not exploiting the world, but living in the world, contained by its intimately rounded spaces.

These shells are more than stunning and abstract geometries. They are animal architecture, animal houses. There is a kind of wonderful alchemy in these creatures that are able to turn so much mud and water into their own houses. One contemporary French philosopher has written a kind of philosophy of shells. In a book called *The Poetics of Space*, Gaston Bachelard writes poetically of the kinds of spaces that attach us to the world, and of the images through which we learn to see the world as a place we might live in. Nests and shells, writes Bachelard, have been particularly powerful in this regard in educating the human imagination. About shells, he says:

> For here too, as with nests, enduring interest should begin with the original amazement of a naïve observer. Is it possible for a creature to remain alive inside stone, inside this piece of stone? Amazement of this kind is rarely felt twice. . . . But an empty shell, like an empty nest, invites day-dreams of refuge. (Bachelard 1964, 107)

The shell is the shelter for the animal. Bachelard calls the "shell-house" a primal image. Shells are extraordinarily simple, which is the source of their power in the imagination—in what he calls the human "dream of inhabiting" the world. Thus he says with sweet understatement, "I believe it is worthwhile to propose the phenomenology of the inhabited shell" (Bachelard 1964, 107).

These shells in the delta's mudflats have the power to take hold of the imagination. Something about their simplicity suggests that the value of this landscape can be seen in its humble clam. The telling image becomes a way in which the value of the place is condensed and delivered to our imaginations. This largely forgotten, antiplace of the delta, if you will, has been treated in largely negative rhetoric. It desperately needs images that recreate it in our minds—the prelude to recreating this discarded place in fact.

The dream of clamshells is a dream of inhabiting the world, of making a home in it, according to Bachelard.

That evening Karl makes a similar point in a much funnier way. While Bernd builds an enormous fire in a dugout hole in the clamshells, Karl recites a song that folksinger Pete Seeger had made popular some time ago. It tells of a man who came to the Pacific Northwest looking for his fortune in timber and fishing and gold. Each failed him. Finally, he finds his fortune in the clam beds of Puget Sound. The homey clams change him:

> *No longer the slave of ambition*
> *I laugh at the world and its shams*
> *And think of my happy condition*
> *Surrounded by acres of clams.*
> (Pete Seeger, "Acres of Clams")

That night I again climb into my sleeping bag and I lie down myself, surrounded by acres of clams. I'm cozy in my bag, "happy as a clam." I might as well be dreaming of clams, so powerful had the image become in my mind. Sometime after midnight, I am awakened by the sounds of all these millions of clams.

When I wake up, at first I think I'm still dreaming. I can't tell what woke me. Then I realize it's a noise outside the tent, though I can't tell what it is. I stumble out of the tent, onto the clamshells, and discover that Karl is already up.

I find I have walked into a beautiful, waking dream. These little islands of shells we're now camped on are lower than Isla Sacatosa, our first stop. We are on one dot of island amid a little constellation of shelly islands and beaches that, I am told, are called *Las Isletas*—the tiny islands. They feel more ephemeral, more fragile even than Isla Sacatosa.

Karl is watching the seas. The tide is rushing in violently. The moon above is full and tonight, he says, we are going to have one of the highest tides of the year. "It's going to be almost a twenty-foot tide," Karl says. "I hope I've calculated right."

He is worried that the tidal waters will crest over the top of the chenier and flood us out. He says he worked out the math, but he is clearly nervous. The sea has swirled in all around us. We are completely stranded on the shelly island and the water has risen to well within a foot of the top of the part of the island where we have camped. If he calculated wrong, we are in real trouble.

As it turns out, we are safe. Despite the rush of excitement, the sea is running as high as it will get. And then I realize that the whole world around us is

luminous. The high-tide moon is full and bright. The clouds have cleared. The moonlight reflects off the seawater, surging in waves. The shells of the island itself are glowing in the lunar light.

The seawater sweeping through the clamshells is making them rattle loudly—one is tempted to say, "clamorously." This was the noise that woke me up. The music of clams. And then there were the shorebirds. As the waters drive them inland, they have congregated in the distance in huge winter flocks. They are all pumped in the night, feeding and screaming.

I am awed by the moment. It is like being in that magic state between sleeping and waking. The delta seems suddenly reanimated and reenchanted. At moments like this, a place enters the heart. With each wave, the shells pound into a loud crescendo and then rattle into a soft whisper as the waters recede. The shorebirds are calling. We seem to be sailing on a luminous island of shells through a darkly lit sea.

*Long-billed curlews wade in the shallows of Las Isletas.*

# SIX

# A River of Trees

MEXICAN "RARE" SPECIES: Great blue heron
PLACE: Riparian corridor, Colorado River

WE ARE STANDING in another of the magical places in the delta, just below the newly built San Felipito Bridge. For many people who have come to know the delta, this spot is their favorite. I am with Carlos Valdés Casillas on a beautiful day in late November, when the heat of the delta summer has eased and it is glorious to be exploring the region. In the emerging geography of Carlos's delta, a conservationist's delta, this is a special place.

Our ground is a bank of rocks, probably purposefully tumbled here to contain the river in its floods. Now the banks hold a pond of green water and the pond is like a magnet, attracting both trees and birds. Several ducks float on the pond. On a far rock is a great blue heron, fishing. In this part of Mexico, great blue herons are rare. They are officially listed as "rare" in Mexico. All around the pond is a forest of cottonwood trees and huge willows.

"This patch of trees," Carlos explains, "is the single best cottonwood-willow forest on the entire lower Colorado River. It's three thousand acres in size." Carlos loves the marshy Ciénega de Santa Clara. But like many others who have explored the delta and encountered some of its hidden jewels, he feels a special connection to this place. At a time when most people had proclaimed the delta sucked dry and beyond hope, Carlos was poking around, looking for habitats. What he found was this forest.

Here he discovered that the stretch of Mexico's delta through which the Colorado River passes en route to the sea has also grown up in the last twenty years. The sinuous green forest along the banks of the river has become so

◄ Surrounded by ditches and farmlands, the Colorado River still manages to flow through the Mexican delta, thanks to abundant rains during El Niño years. Cottonwoods and willows follow the river of trees.

▲ Endangered in Mexico, the great blue heron finds a haven in the ponds near the river through the delta.

extensive, in fact, that it puts similar habitat in the United States to shame. Cottonwood and willow trees grow in well-watered places. Their forests are called riparian habitat. Once the entire lower Colorado River, from the Grand Canyon south to the Gulf of California, was a long linear oasis, a green belt of riparian forests.

In the 1970s a number of biologists began to study the importance of riparian habitat for wildlife in the American Southwest. They quickly realized that these groves of cottonwoods and willows are an invaluable resource. Because these forests grow where their roots have access to freshwater, their green against the desert brown is a sign of water—for people, they signify relief from the desert and life itself. Biologists' surveys discovered that bird densities in cottonwood groves are literally hundreds of times higher than in the surrounding desert. One study showed that as many as one thousand pairs of birds breed in as little as one-hundred acres of cottonwoods. The studies have also shown that riparian and wetland habitats support by far the greatest biodiversity of birds in the entire Southwest.

As Carlos and I stand on the rocks, we look into the forest. Birds move through the branches, the leaves of the cottonwoods rustle in a soft wind. In the arid Southwest, these groves are the focus of life.

What is remarkable is that the Colorado River delta should have such a large stand of cottonwoods. Carlos not only discovered they were here; he discovered that the cottonwood-willow forests in the Mexican delta are much more extensive than any of the same habitat in the United States. From the Grand Canyon to the Northern International Boundary (NIB) between Arizona and Mexico, the stretch of the Colorado River in the United States is five times the length of the river through Mexico to the Gulf. Yet Mexico has many times the riparian habitat in the delta than does the much longer stretch of the lower Colorado in the United States.

Like the Ciénega, the riparian corridor was created by accident, another unintentional windfall from the United States management of the river. The river of trees was created out of "wastewater," as well, though of an entirely different kind than that found in the Ciénega. Like the Ciénega, this accidentally created riparian zone again demonstrates how the delta has been treated as a kind of dumping ground for the "plumbing system" on the United States' portion of the river, where little attention has been paid to the consequences for good or ill south of the border.

When the United States' system is flushed, everything flows down to the bottom of the river. This forest takes us into the heart of *la problematica del delta*— the problematics of the delta. In a word, water.

To understand how a forest of this size and quality could exist in the delta, land that is supposed to be lifeless, I have come with Carlos Valdés. I could have chosen no one better, the man who has become the delta's greatest champion inside Mexico. At the time he found this area, Carlos was a professor and researcher in biology and ecology at the Instituto Tecnológico y de Estudios Superiores de Monterrey (ITESM), Guaymas. Until recently, when he took a major new position in Canada, Carlos lived in Guaymas, a beautiful town halfway down the eastern coastline of the Gulf of Mexico. At ITESM, Guaymas, he was the undergraduate professor of Osvel Hinojosa, the student working on rails, and Jaqueline García, the Ph.D. candidate working on ecotoxicology. His close relationship to Ed Glenn, the University of Arizona's Environmental Research Laboratory botanist who has done so much work on the Ciénega, is illustrated by the fact that these young Mexican students went from Guaymas to Tucson to study under Ed, as have several other students. As close friends and colleagues, Carlos and Ed have been the main forces behind the conservation emphasis in the delta. Many people, including Ed himself, credit Carlos with being the person most responsible for changing our perception of the delta. Carlos is a jolly person with a dark beard and a Tommy Lasorda belly. He radiates irresistible Mexican warmth—*simpatico*.

In characteristic fashion, when I tell him that Ed Glenn views his work as central to the revival of interest in the delta, he deflects the credit back to Ed.

Carlos grew up in Mexico City. As an undergraduate in the city, he studied under Enrique Beltran, one of the early legendary figures in Mexican conservation. Enrique was a friend, for example, of Aldo Leopold and his son, Starker. Later, Carlos earned a Ph.D. in biology from Oregon State University in the 1980s, and it was there, reading Aldo Leopold, that he began to develop an interest in the delta. His professor asked him about water issues in Mexico. That led him to read Aldo Leopold's "Green Lagoons," which moved him deeply. When he returned to Mexico, Leopold's accounts inspired Carlos to research the delta.

At first, Carlos interviewed people in the delta, learning from them that the delta was not dead as had been reported. Ed Glenn got in touch with him in the early 1990s, and together they decided to focus their research, to understand what was happening in the delta. Carlos began to assemble an inventory of delta habitats in the 1980s and through the 1990s. Ed then joined Carlos in drawing up a new map of the natural habitats of the delta. In addition to interviews, Carlos consulted satellite and aerial images. Besides the Ciénega, he discovered that a number of new habitats seemed almost miraculously to have sprung up south of the international border.

"We found that the Mexican part of the delta contains thousands of acres of cottonwood-willow forests," he says to me, as we look at the trees. "These habitats are rare north of the border in the United States."

Carlos has divided the riparian habitat along the river into six zones, going from north to south. There is a big section of cottonwood and willow forest along the river as it separates Arizona from Mexico, just below Yuma to San Luís. This area is called the limitrophe, which specifically refers to the twenty-four-mile stretch between the Northern and Southern International Boundaries, where the river becomes the boundary between Arizona and Mexico. It is an undeveloped stretch of habitat. In many places, the cottonwoods are merely a thin ribbon along the much-shrunk Colorado, which winds through thousands of acres of farmland. But just below the levees, near the San Felipito Bridge across the river, the thin ribbon bulges into this big beautiful forest—a remnant of the fabled forests that once lined the river. A riparian gallery forest contains a variety of trees and shrub species, of varying heights, including tall cottonwoods and willows.

Between the Ciénega de Santa Clara, some of the best wetlands on the continent, and this unrivaled riparian forest that lines the river in Mexico, the delta is home to some of the richest habitat on the lower Colorado River. It has inspired the biologists who study it, as well as the people who live here, to dream of preservation and ecological restoration. The irony is sweet: the Mexican delta, dead or invisible to the minds and hearts of Americans, has a greater potential for restoration than do the stretches of the river in the United States.

"This is really good habitat," Carlos says. "It's a partial restoration of the riparian vegetation, including native species, in the Colorado River delta in Mexico."

How this riparian forest in the delta returned is yet another of those great and unlikely stories of the delta. Carlos and Ed solved this natural paradox—the return of the delta forest in a land of the dead—and in the process demonstrated a crucial link between the growth and health of these forests and freshwater in the river.

Carlos pauses. He looks into the forest that has inspired his professional life. A breeze stirs the cottonwoods. It is a forest to stir the heart, just as the wind stirs the trees' heart-shaped leaves.

We think of trees as stationary, rooted in place. They are not supposed to travel, not supposed to come and go. That is why Macbeth, for example, thought the witches' prophesy left him invulnerable when he stole the crown of Scotland: He would not be conquered until Birnan wood moved to Dunsinane, the witches told him. A moving forest is an impossibility, a miracle, but not along this river of trees.

If forests' immobility were a reality, then this forest along the river in the delta is a miracle. What's more, it's an everyday miracle. The cottonwoods and willows in the delta are a moveable forest, visible and living markers of the vicissitudes of the ecological health of the delta as they have come and gone through the century. And for the riparian forests in the delta, as for nearly everything else in this desert, the central fact of life is the same: The trees come and go with the water. They ride in on floods as floating seeds and die off in droughts.

Carlos explains that the delta was nearly devastated when the dams in the United States were built during the last century. They cut off water to the delta and the huge forests of cottonwoods dried up and died. Estimates of the original forest's size along the river in the delta vary. Although incidental descriptions of the bottomland cottonwoods have come down to us from early explorers, no one did baseline botanical studies before the dams went in. So giving a precise number is difficult. Early travelers through the delta were amazed by the size of the woodlands, which became legendary in their reports. Perhaps the best indication we have of the original extent of these forests comes from James O. Pattie, the same hunter and explorer who corroborated Aldo Leopold's belief that jaguars roamed the marshes and forests. In 1827–28, Pattie canoed down the delta, trapping beavers in the "lofty and thick set forests," which frequently flooded:

> The river below the junction with the Helay [Gila], is from 2 to
> 300 yards wide, with high banks, that have dilapidated by falling in.
> Its course is west, and its timber chiefly cotton-wood, which in the
> bottoms is lofty and thick set. The bottoms are six to ten miles wide.
> The soil is black and mixed with sand, though the bottoms are subject
> to inundation in the flush waters of June. (Pattie [1833] 1966, 142)

If the wetlands were between six and ten miles wide in the bottomlands south of what is now Yuma (the junction of the Gila and Colorado), they would have extended about seventy-five miles down toward the sea, where they would have reached the zone affected by the salt tides.

That means there would have been about 750 square miles of wetland forest with marshes. This huge green belt through the desert—bordered abruptly by desert—would have been nearly half the size of our Grand Canyon National Park. If only half of that had been forests, that means 375 square miles—or 240,000 acres—along the sides of the Colorado River were cottonwoods and willows. Also surrounding this river through the desert were large stands of arrow-weed (cachinilla) and mesquite forests, fringing out into the desert.

The huge forests of cottonwoods and willows depended on fresh water. Their existence inextricably linked to the flows of water in the river, they began to shrink in numbers quickly when the dams arrived. Since 1935, when Hoover Dam came on line, these flows have been graphed. For several years after 1935,

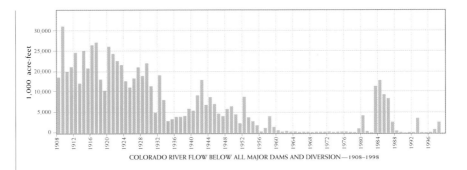

COLORADO RIVER FLOW BELOW ALL MAJOR DAMS AND DIVERSION—1908-1998

as the reservoir behind Hoover Dam filled and formed Lake Mead, very little water flowed to the delta. That's when the delta began to dry up, when the clams in the upper Gulf began to die and the trees and marshes along the river began to vanish.

There were small releases of water to the delta in the 1940s and 1950s, ranging from about one-half to one-third the pre-dam flows. Some years, little to no water reached the delta or the sea. By the early 1960s, Lake Powell was filling behind the Glenn Canyon Dam. This was the period in which the delta got its current reputation as a dead wasteland. The 1944 water treaty with Mexico calls for about 10 percent of the base flow of the lower Colorado River to be delivered during nonflood years. This 10 percent of the river's water technically reaches the delta, but not the delta wildlands. The Mexican allocation is diverted at the border into the Canal Central in Mexico for irrigation in the Mexicali and San Luís agricultural valleys. In "normal" years, then, no water reaches the wild parts of the delta below the farm levees.

A graph of the water flows since 1935 shows that for nearly forty-five years, very little water reached the delta. By 1964 Glenn Canyon Dam was functional. All water in the river was captured behind the dam. According to Ed Glenn, the biologist who has studied the wetlands of the delta so carefully, "Lake Powell filled sometime around 1981."

During that time, scarcely a drop of water reached the delta from the river. Then El Niño struck. The Colorado River entered a wet cycle. Unexpected amounts of rain and snow poured into the river. "The big 1983 floods were the beginning of a new era of releases with the El Niño cycles," Ed says. Enormous releases of water spilled out of the dams and roared into the delta.

During several of these years, furious floods swamped the delta farmlands and wetlands. Through the floods of 1982–1983, the delta received about two and a half times its normal treaty allocation of 1.7 million acre feet (Mary Orton, "Restoration Opportunities on the Colorado River Delta," presentation at conference "The Law of the River," Loews Ventana Canyon Resort, Tucson, May 4, 2000). A local historian named Oscar Sanchéz told me the floods in the delta nearly destroyed family farmers. It was, he said, *"un burro tremendo"*—a huge wall of water—that came down upon them.

While the floods have been less through the 1990s, three spring releases—in 1993, 1997, and 1999—again sent large quantities of water into the delta, big spikes of water. In ten of the eighteen years after 1980, the water in the river was too great for the storage capacity of the dams. With peak floods sending almost 3.9 million acre feet some years into the delta, the region received almost three times the normal treaty allocation.

At precisely the time when many American writers were describing the delta as dead and buried, parts of it were greening into a spectacular delta oasis, a salted-out desert with huge patches of green in the Ciénega and the riparian habitat along the Mexican reaches of the river.

Shortly after the floods of the early 1980s, satellite images revealed that about 100,000 acres of riparian habitat, distinct from the marshes of the Ciénega, had returned to the Mexican portion of the delta. Since those initial floods in the 1980s, the delta has not sustained that level of riparian habitat. The greenwoods in the delta have shrunk to about 6,500 acres. Still, that's much larger than in the lower Colorado River in the United States.

In fact, there is nothing like it along the lower Colorado River in the United States, where it is both much harder and much more expensive to try to restore riparian habitat. Also, the experiments to do so have not been particularly successful. In the American portion of the lower river in the 275 miles south of Davis Dam, surveys in the 1970s and 1980s by U.S. biologist Robert Ohmart found less than 500 acres of cottonwood habitat, which he described as "scattered groves containing a few mature individuals" (Ohmart et al. 1977, 45).

The tens of thousands of acres of riparian habitat in the Mexican delta in the 1980s were not even half of the original extent of the riparian corridor. Still, these forests were large by any standards in the ecologically abused Southwest.

The return of the cottonwoods along the river in the delta offers a message, simple and clear. It is the same message found in the accidental birth of the great marsh, the Ciénega de Santa Clara. The delta can be restored. All that is required is water. No expensive restoration projects. No grand designs. As scientist after scientist has told me, in the delta all that needed to be added was freshwater. Nature did the rest.

There is clearly enormous untapped potential for ecological restoration in the delta, more so than in the United States. Carlos is very optimistic about building on and expanding what nature has begun. "This is the revival in the delta," says Carlos. "It's proof that we should be facilitators of the environment."

"I'm from Mexico City," he concludes. "I know about abused environments. How many environments are more abused than the delta? Yet we give it a little water and it comes back."

*Sam Spiller of the U.S. Fish and Wildlife Service gestures at the new trees on the Pratt Farms Cottonwood Restoration project near Yuma, Arizona. Restoring riparian habitat in the U.S. portion of the lower Colorado has proved difficult.*

The great part of this revival in the delta is that nature led the way. The unfortunate part of the revival is that it was accidental and could be easily reversed. It happened because El Niño floods swamped the dams and forced releases that no one expected. The accidental results are positive, but we must learn from them and build upon these lessons of restoration.

"It's a kind of unconscious mitigation process," Carlos says, using the technical language of environmental law and conservation. By that he means that the managers of the dams in the United States have by their actions done something to "mitigate" or compensate for the environmental destruction in the delta caused by the operation of the dams upriver. But they had never intended to do anything restorative.

The return of rich habitats in both the Ciénega and the riparian corridor along the river prove that ecological restoration in the delta is both desirable and, more crucial, doable. The problem is, both revivals resulted from accidents—one historical and human, in the case of the Ciénega, the other natural and climatological, in the case of the riparian corridor. In both cases, though, revival was directly related to management actions taken upriver in the United States, sending water down into the Mexican delta. The question raised by both the Ciénega and the riparian corridor is simple: How can we guarantee and sustain these accidental revivals?

Scientifically, that means determining how much water is required to support the ecological restoration in the delta. Ed Glenn has taken the lead in addressing this issue. He studied all the El Niño flood pulses of water flowing into the Mexican delta during the 1990s—a small release of 200,000 acre feet in 1997, a large release in 1998–1999 of 2.5 million acre feet.

Then he analyzed the patterns, concluding that very little freshwater released into the channel would be required to sustain the delta's tenuous revival. One larger simulated flood release of about 250,000 acre feet every four years will enable the cottonwoods and willows to germinate. Then sustained flows of about 30,000 acre feet every year will keep them alive.

"Water controlled the return of cottonwoods and willows to the delta," Ed says. "We can guarantee their survival with less than one percent of the total flow of the river." The water for the Ciénega, another 120,000 acre feet, is a separate matter.

Ed's research gave rise to a campaign along the lower Colorado River called "1 Percent for the Delta." What has startled people is how little water is required. Finding ways to persuade water managers on both sides of the border to make the environment a priority has been difficult. Nevertheless, this research has provided a tool: we have a concrete quantity of water required for ecosystem regeneration in the Mexican delta—the richest part of the lower Colorado River.

"It's a specific quantity and it's reasonable," Ed says. "It advances the discussion."

The accidental revivals in the delta have inspired conservationists to work for a guaranteed revival of the delta. Maybe later we can consider an expanded revival. The trick or the challenge is to create binational mechanisms for ensuring the necessary water. That will require convincing water managers in both the United States and Mexico to dedicate water in order to sustain the revival. Convincing these managers, "water hogs" or "water buffaloes," as they are often called (even by themselves), is a tough sell. In Mexico managers are only just becoming accustomed to thinking in environmental as well as developmental terms. In the United States, the "hogs" have consistently resisted.

More fundamentally, though, nature has taken the lead in yet another way in these revivals, by demonstrating that the two sides of the border are not as separate and distinct as we have liked to imagine. Realizing that the U.S. and Mexican deltas are two parts of a single ecological whole is perhaps the first step in understanding that we need to manage the lower Colorado River and the delta as a unified whole.

As Carlos puts it, "The forest here in Mexico is an essential part of the lower Colorado riparian zone. The delta on both sides of the border needs to be managed as a single unit. Any scientific plan for the Colorado River must include the delta."

This is the heart of the conservationists' dream for the delta—to manage it as a unified entity. "We are linking the borders with our research," Carlos says. The green thread of the linear oasis helps stitch together the two sides of the border, part of a new and expanded vision of life in the Southwest. Conservation across borders is central to restoring the delta, and its foundations must be not only scientific, but also ethical and political. Crossing borders and finding common ground is part of the new human vision along the border, as well as a natural vision.

As Carlos and I prepare to leave the bulging forest around the pond of water nearby, the blue heron lifts on heavy wings and flies into one of the trees. Great blue herons are large birds. This one sits in the tree with its long neck and long legs, gangly and awkward looking. It is almost comical to see big herons in trees. With its big body and neck, it reminds me a bit of a canister vacuum cleaner in the branches.

Overhead a red-tailed hawk circles on the day's thermal currents, and a vermilion flycatcher snatches bugs on short sallies from an exposed branch on a cottonwood. Vermilion flycatchers are one of the most beautiful birds on the continent, with brilliant red bodies like candied apples and contrasting black wings. More powerfully than scientists and politicians, the birds testify to the importance of this riparian habitat as a source and center for life in the desert and delta.

SEVEN

# Crossing the River

ENDANGERED SPECIES: Southwestern willow flycatcher
MIGRATORY SPECIES: Yellow warbler
PLACE: Riparian corridor and the limitrophe, Colorado River

THE SOLDIERS JUMP to their feet and grab machine guns. Two army-green Humvees block the dirt road. We have just stumbled onto a squad of Mexican soldiers camped on this dusty back road along the Colorado River. Not yet sunrise, most of them are still lying on their blankets, spread on the road, when we wake them up and send them scrambling into confused action.

The officer in charge, a lieutenant, grabs his Uzi and walks up to the driver side of our white pick-up. Several of the soldiers gather round us. They wear heavy dark uniforms, a bit wrinkled from having been slept in. The soldiers look at least as surprised as we do, and a lot scarier. We had thought we were out here to study songbirds. Instead, it suddenly looks as if we have driven into an embattled frontier zone.

The lieutenant asks us in Spanish what we are doing.

Michelle Rogne is driving the car. She is in her early twenties, blond, from Wisconsin. She is a field biologist in charge of the team. John Cornell sits in the back seat with me. He is a biologist from Oxford, England. Miriam Lara Flores, a young Mexican biologist from Ensenada, has joined the research team as a representative from the biosphere reserve in Mexico. She and I are already friends from working together on the Yuma clapper rail survey in the Ciénega de Santa Clara. These biologists are conducting a study organized by John Hart and Charles Van Riper, of the U.S. Geological Survey in Flagstaff, Arizona, on migratory songbirds along the lower Colorado River.

◄ *John Cornell holds up a common yellowthroat warbler, part of a survey of migratory birds along the lower Colorado Delta. Note the willow trees in the background and the heavy sweat on his clothes, though it is still early morning.*

▲ *Flycatchers like this black phoebe are common in the woodlands of the delta.*

123

Michelle does not say a word. She just hands him our paperwork and permits. The lieutenant studies the papers while the soldiers eye us suspiciously.

The scene is a reminder that there is trouble along the border with Mexico. These soldiers are here as part of the international war on drugs. Some of the heaviest drug traffic along the entire United States–Mexico border passes through the delta of the Colorado River. All the major roads in the delta have military checkpoints on them. You have to get out of your vehicle while soldiers ask you questions and search for drugs or guns. It is a common experience for any traveler in the delta. A gringo can automatically arouse suspicions.

Still, none of us in the pick-up truck has ever run into a group of soldiers like this. The situation is a dramatic reminder that ecological issues are never too far removed from other social and political issues in the delta. Along this border, the problems have proved especially difficult to control. The United States has focused on controlling the border for drugs and illegal immigrants. Increasingly we are discovering that the health of the environment is itself a social problem, part and parcel of the rest. This border is more porous, more fluid as it were, than we have imagined.

As researchers, we are concerned about migration, too. Unlike the soldiers, though, we are not interested in illegal human movements across the border. We are interested in animal migrations up and down the continent and across the border. So the moment has a strange resonance for us, as ecological and social issues parallel each other in a tense moment in the field. We normally think of nature and culture as separate, even conflicting. But that's an oversimplified version of things. Nature reflects culture, and the other way around. Along the frontier, nature is a crucible in which you can find every social issue played out.

After the officer studies our documents, he leans into the window of the pick-up and looks us over carefully. A smile opens across his face. He asks us exactly what we are doing here. Miriam answers him in Spanish. She explains that we are going up the road a bit, and then turning into the forest along the river, where we are going to set up nets and catch songbirds.

The soldier looks bemused. Bird-watching in the woods strikes him as quaint and harmless next to the more high-profile and dramatic battles going on along the frontier of drug smuggling and illegal immigrants.

Yet like virtually all the studies I have been on in the delta, this study has potentially huge political implications. These go to the heart of the water issues between the United States and Mexico and are important in understanding the way water is allocated along the border—among political entities as well as between people and nature. In this part of the continent, no single issue may be more contentious than water, though drugs and illegals get most of the press. They seem more dramatic, the stuff of guns and raids and subterfuge—as in the movie *Traffic*. Yet water plays a crucial underlying role in border stability and is a key to both environmental health and social prosperity.

The lieutenant lets us go. Two big military Humvees fire up and edge to the side of the road. We drive past, relieved, laughing. Then we turn off into one of the most beautiful forests on the entire lower Colorado River.

After the encounter with the soldiers, we enter a shaded grove of cottonwoods and willows. We are very close to the spot where Carlos and I had visited, arguably the best riparian forests on the lower Colorado. It feels like a refuge from the real world. Some of the trees are fifty feet tall. In the soft breeze of the morning, the cottonwoods' leaves flutter beautifully. The ground is a litter of leaves and old branches. Local Mexicans often come here to collect firewood for their homes—for cooking and for heating on cold nights.

The birds are already singing this morning, early in their fall migration. We quickly swing into action, setting up the nets we will use to try to capture some of them. The delta and the waterway up the Colorado River are thought to offer a route for migrating birds in the West. Hart and Van Riper—the organizers of this study we are conducting—have chosen to focus on small songbirds called passerines. These are the often brightly colored songbirds we are all so familiar with in our backyards and in the northern woods in the summer—birds such as warblers, sparrows, and tanagers. In the winter, these songbirds travel south to their wintering grounds in Mexico and Central America, the so-called "new-world tropics," or neotropics. Hence the birds we are researching are called neotropical songbirds.

Considerable research has been done on migratory birds, but most of it has focused on the major East Coast routes. Little is known about the routes along the Pacific Coast.

The nets we use are about forty feet long and ten feet tall and are called "mist nets," because the webbing is fine as a mist. They are suspended between thin black metal poles, which are staked into the ground about every ten feet on the nets. The result looks vaguely like very delicate volleyball nets set up in the forest, except that the net comes all the way down to the ground. Birds don't see the nets, fly into them as they forage low to the ground, and become entangled. That way the biologists can capture the delicate little creatures, weigh and study them, and put tiny leg rings on them—keys to tracking migration patterns.

We had intended to put the nets up in exactly the same place the team had put them the previous spring. But we can't—the flooding river water pushes us out of the some of the old locations. So we improvise, lining up six mist nets under the big cottonwoods in the forest. Then we wait, resting on the tarps we have spread over the leafy ground. Every ten minutes, we check the nets to see what birds might have enmeshed themselves. Though only about 8 A.M. by this time, it is still hot enough to work up a sweat without moving. We will need to finish this work by about 10 A.M., at which point it will be too hot to move about. We're on our way to well over 110 degrees Fahrenheit today, even though it is fall migration time. The shade of the trees offers sweet relief.

I love the experience of checking the nets to see if any birds have been caught. There is a sense of keen expectation. Then when you see that there is a bird in the net, you have the rush of curiosity and wonder—what species might it be? I have mist netted often in the past, helping capture birds that brought through amazing surprises, species that turned out to be new records for the entire area.

When we check the nets the first time, we have captured two small birds. One is a little flycatcher, a Trail's flycatcher. These birds are important insect eaters and are closely related to one of the premier endangered species in the area, the southwestern willow flycatcher. The other bird is called a common yellowthroat warbler. Warblers are a whole family of small songbirds. They are brightly colored and, as their name implies, have some of the most wonderful songs of all the birds of North America. The common yellowthroat warbler (*Geothlypis trichas*) is one of the best known of all. It's common in wetlands of all sorts, a brilliant yellow bird with a distinctive black-bandit mask across its face. Its *witchedy-witchedy-wooo* song is easily recognized and deeply beloved.

The sweet rush of finding a bird is exceeded only by the joy of holding these delicate creatures in your hands. You have to use special techniques to disentangle them from the net and hold them in your hand. And then you have the pleasure of seeing them with breathtaking closeness.

We process this yellowthroat, weighing it, measuring it, and giving it a leg band so its movements can be tracked by other bird-watchers. The yellowthroat weighs almost nothing. It is little more in my hand than a tiny bundle of feathers with two dark gleaming eyes. Then we release it. I hold it up between my fingers, say good-bye, and open my palms. The warbler looks at me, realizes that it is free, and flies away. I love that gesture of letting go—and the bird's quick flight to freedom as it returns to the woods.

At 9:10 in the morning, we capture another small warbler. This one is harder to identify. Like many species of warblers, this bird has lost its distinctive breeding plumage during migration and has become a much more nondescript creature of olive greens and pale yellows. It has a faint white ring around its eye. Michelle holds it gently in her hand and opens a copy of Peter Pyle's amazing book, *Identification Guide to North American Passerines: A Compendium of Information on Identifying, Aging, and Sexing Passerines in the Hand.*

The subtitle could not be more descriptive of the book. This book is like an unabridged guide to identification of passerines. It gives pictures and facts, not just at the species level, but detailed accounts of every race of a species and all the plumages for every age of the species. Michelle studies the images of warblers in Pyle's book and narrows the bird down to a yellow warbler (*Dendroica petechia*).

Measuring various feathers, particularly the rectices or outside tail feathers, Michelle identifies the bird with amazing precision: a male in its second year.

"It's one of our target species," she says. "It's certainly migratory."

The researchers are particularly interested in about seventeen species of passerines—especially warblers, sparrows, and flycatchers—as indicators of movements up and down the Colorado. It is astonishing, actually, how little is known about the migratory patterns of these birds in the West. The patterns are part of a complex and invisible animal geography that crisscrosses the continent. In these birds' cases, the geography is made up of a web of unknown flight lines that stitches the continent together.

Of the approximately 650 breeding birds in the United States, only about 100 species are completely nonmigratory. A huge proportion of the birds that do migrate head to the tropics in Mexico and Central America—more than 200 species. It is in the context of these invisible geographies that human lives are located. As Scott Weidensaul puts it in *Living on the Wind: Across the Hemisphere with Migratory Birds*, "Migratory birds as a whole are being squeezed at every stage of their life cycle—summer, winter, and in transit" (Weidensaul 1999, 344). Like so much else in the delta, these songbirds have their own geography, and it is one that defies the two political jurisdictions of the United States and Mexico.

Daniel Anderson from the University of California, San Diego, is an expert on white pelicans. He has studied them throughout the West and in Mexico. They have been in long-term decline since the 1920s. These and other birds that summer in the Klamath Basin, in southern Oregon, and in northern California also winter in the delta. He describes these birds as a "synthesis of the problems in the delta."

Anderson has identified the delta wetlands as crucial parts of the Pacific Flyway for migratory birds. Protecting habitat for migratory birds is a major international challenge. It is not enough simply to try to preserve habitat in the United States, for example. We must also work jointly, in this case with Mexico, to preserve habitat on both sides of the border.

Part of the purpose of the study of the migratory songbirds is to try to understand their relationship to the riparian habitat along the Colorado in the delta. No conclusions have yet been reached. The study is still too young. But one thing is clear. To protect species of songbirds, we must protect and encourage these cottonwood-willow stands.

This morning in the cottonwood grove, we capture seven birds. That number is down from the previous spring. But it is much better than the number captured by the other group in the team of biologists on this project. They were working nearby in what is now the most abundant type of forest in much of the lower river—salt cedar, or tamarisk. This ugly tree was introduced in the 1880s and has since nearly taken over the lower river. It is an invasive plant, sucking up large amounts of water and leaving marginal and salty soils. No birds were captured in the salt cedar this same morning. About 80 percent of the floodplain of the current lower delta is now covered by these weedy tamarisk, which are nearly impossible to root out and eliminate.

After identifying the yellow warbler, we get ready to let it go. Michelle hands it to Miriam, who takes it carefully between her fingers in a gesture of infinite tenderness. Since birds are creatures of flight, it seems somehow startling to be holding them in the hand. They always hint at something transcendent, a flight of angels. I know that in this time in which nature is increasingly endangered, we are usually told we should maintain a hands-off attitude toward nature. But I cherish moments of close encounters with wild creatures. It is one of the reasons I love studying wild creatures with scientists—for the opportunity to know the animals face to face. Close encounters with wildlife have largely vanished from our lives.

Miriam holds the warbler out in her hand. Like the trees themselves and the prosperity of people, it is really water that draws the songbirds here. Even more than heat, water is the controlling factor in the delta. Everything in the delta—wealth and poverty, happiness and misery, desire and disappointment, anger and salty bitterness—seems to depend ultimately on water. Or its absence.

Miriam laughs as she lets the warbler go. It disappears into the cottonwoods, where it resumes its migration through the delta. Watching the warbler fly off, I remember something Aldo Leopold wrote in *A Sand County Almanac*. "Hemispheric solidarity is new among statesmen, but not among feathered navies of the skies" (Leopold 1949, 35 ).

I'm sitting in the cool comfort of an air-conditioned museum in Mexicali, the Museo Universitario de la Universidad Autónoma de Baja California, Mexicali. I'm looking at small black-and-white photos from the nineteenth and early twentieth centuries along the river. I have come here because I want to try to understand what the legendary forests of native riparian habitat must once have looked like along the river—how large the forests were, their appearance, their relation to the river itself.

This museum is another kind of shaded grove, not the leafy green of cottonwoods, but a kind of academic retreat into the leafy pages of files and books and pictures. Outside, a stream of cars flows round the building. In a sense, Mexicali has made the river flow through a forest of buildings and roads. It is one of the fastest growing cities in Mexico, a town straddling the border with California. Only a century ago, almost no one lived in Mexicali. Its population was too small in 1900 to be included in the area's official census. In 1910, its population was given officially as 462 hardy desert people. Now it is bursting past 1 million, with a growth rate of about 3 percent. Once it was a cow and cotton town. Today it is an international *maquiladora* city. The Colorado River sustains the city's growth. Like the cities on the U.S. side of the border, Mexicali also has a nearly insatiable thirst, another big-gulp city in the desert.

With me in the museum are Gina Walther, the director, and Alberto Tapia Landeros, a university researcher and writer on local natural history. Alberto also writes a column on natural history topics for the local newspaper. Alberto has spent his entire life in the delta. His writing is an effort to comprehend and reconstruct the former delta.

The museum has a collection of about 3,000 photos of early life in Baja California. About 150 of them are from the delta a century ago. They are an invaluable resource in trying to imagine what the delta was. Several of the small and out-of-focus photographs in the museum collection are particularly striking. One of the photographs, for example, shows a young native Cucapá man in long hair, long-sleeved shirt, long pants tied at the waist with a cloth belt, and bare feet. He is standing on the bow of a boat on the Colorado River. He is a workman on the riverboats, earning money in the early twentieth century. In his hands he holds a long pole, like the one Juan Butrón uses, with which he propels the boat along the river.

Beyond the Cucapá man is a staggering background. All along the far side of the river, a dark line of tall trees defines a forest that is both dense and wide. The point about the image is that its dimensions are so large as to be undefined by the photo. I see a cottonwood and willow forest along the bank of the river. With this image, the full extent of the pre-dam forest comes home. The little thread of cottonwoods along the modern Colorado in the delta is a faint echo of this former *bosque*, or forest. Today there are groves where there was once forest.

Alberto describes the original ecosystem along the river as *"un hilo verde,"* a green thread through the desert. *"Ese hilo eran bosques verdaderos de sauces y álamos junto a las aguas."* That thread was made up of true forests of willows and cottonwoods along the river. He is referring to the hundreds of thousands of acres of riparian habitat in the delta prior to the upriver dams. What Alberto offers is something beyond statistics, though. He offers a Mexican perspective on the value of these forests—their richness for creatures and people.

*"Esta cuenca era un paraíso,"* Alberto proclaims. This lowland basin was a paradise. According to his book on the natural environment of the northern Baja, *En El*

*The pre-dam delta was legendary for its vast wetlands and forests, supporting beautiful birds like the snowy egret.*

*Reino de Calafia*, there were thousands of beavers, muskrats, and even river otters in the arms and lagoons of the river. He writes that there were 22 species of bats, both kinds of deer—white tailed and mule—and many other mammals. It was the birds, he says, that took "most advantage" of the Colorado River. He claims there were 450 species of birds in the delta.

According to the botanist who has made a careful study of the plants of the delta, Ezequiel Ezcurra, there were as many as 420 species of vascular plants in the delta wetlands. Periodic floods maintained this whole area. Ezequiel is currently the President of the Instituto Nacional de Ecología in the Mexican government, a botanist who has done some of the most important studies of the plants in the delta. Not only has he documented much of the delta's former abundance, but his work offers one more illustration of the way the Mexican delta is part of a larger ecological whole, linking the United States and Mexico across the international boundary.

Ezequiel says, "Whole species were killed off within ten years after the dams were built." When he ventured into the delta decades ago, hoping to find many of the plants he had studied, he found the opposite "of what my heart expected." He says there have been at least fifty "local extinctions" of plants, maybe as many as two hundred. These include one species of willow, another of reed. We can't know for sure, he emphasizes, because no one compiled an inventory of plant species before the dams were built.

Ezequiel emphasizes another crucial and biologically fascinating point—one that underscores the dynamic interrelation between the delta and the rest of the river. The green thread through the desert actually constitutes a kind of southern extension into Mexico of plants that are typically more northerly species. The extinctions in the delta are local. No complete species have gone extinct; rather, they are disappearing from the southernmost part of their range.

He describes the dynamic relationship between the river and its wetlands. The river transported the trees and plants down into the desert. "The river was always flowing with seeds," he says. "All the flora was brought down into the delta by the river. It was frost-free there. Most of the plants in the delta have a northern affinity. They go all the way up to Canada. There are no plants in the delta, for example, that are from southern Mexico's wetlands. These northern plants in the delta are stunningly maladaptive to life in the desert. But they survived in the wet rich soils."

The Colorado was literally a river of life. It supplied the delta with plants from upstream. In exchange, the waterlogged delta could serve as a kind of nursery for the river forests upstream after periods of prolonged drought. Plants could actually move back upriver, following the river through various means of dispersion. The southern delta thus helped to resupply and recolonize depleted habitats upriver (Fedierici 1998, 14). In other words, the forests along the delta river were not simply grand, extending across boundaries. They existed in a dynamic interrelationship with the river upstream—another feature defining the larger ecosystem of delta and river. The river was a kind of two-way highway for both plants and animals.

It is more than a metaphor to say that these bottomlands were filled with a "moveable forest," coming and going with water. It is literally the case. That is why the forests could return so quickly in the delta with the floods of the 1980s and 1990s.

One photograph created an image that captured the dynamic sense of the old delta and its river. It is a modest photograph, a snapshot really, of a beaver. It is dated March 1, 1901. A single little beaver is crouched on the bare ground. The forests in the delta were once full of beavers. James O. Pattie, the early nineteenth-century trapper and explorer, caught so many beavers cruising down the river that his team had to build new canoes to carry the pelts. The beavers vanished with the forests.

Now the beavers are returning. A biologist who has studied the beaver in the delta, Eric Mellink of Universidad Autónoma de Baja California, Ensenada, described how in the huge floods during the El Niño years, you could see beavers being swept downriver with the water, carried along in the surge of the waves.

Imagining when the Colorado was still a wild river, it might be more accurate for us to think of the delta not as the end of the river, but as the river's throbbing and pulsing heart—the center of the river's massive life. Cut by hundreds of arms, channels, and lagoons, the delta spread out wide and wet across the fertile soils, the same sort of fertility that makes the Imperial Valley in California. Part of the river's one-time delta, covered with deltaic soils, was the most fertile agricultural area in the entire Western Hemisphere.

Most Mexicans I spoke with blame the *anglos* for the destruction of this former delta paradise. It is probably hard for Americans to understand the resentment that many feel toward us. As Alberto puts it in his book, American trappers and other explorers like James O. Pattie began to enter the delta in the 1820s. He writes, *"Desde 1820 los angloamericanos iniciaron una desquiciada cruzada de exterminio. . . ."* (Tapia 1998, 55). Since 1820 Anglo Americans have carried out a crusade of extermination that has unhinged the whole delta.

But the truly intensive development by the *norteamericanos* happened with the dams. Alberto laments the construction of Hoover Dam as the event that destroyed the delta and its life. He says that the *"transformador colonizador"* built the dam and *"provocó aquella sed terrible. Después de 1935, ya nada volvió a ser igual."* The colonizer was the transformer, provoking that terrible thirst (caused in the delta by the dams). After 1935, nothing in the delta was the same again.

In his view, as in the view of almost every Mexican in the delta with whom I have spoken, Americans' role in damming the river gives us a special responsibility in restoring it.

If time is a river, then so is history, especially in the arid Southwest. Water is the fundamental reality of life in this region. Without it, human life here is not possible. The Colorado River flows through the history of the desert Southwest as one of the great truths. We may now live at some distance from that truth, as the river has been transformed into a human construct. But it is still the river that has made human history in the delta region possible.

As Alberto Tapia told me, there is a Mexican proverb that identifies the intimate relation between human history in the delta and the Colorado River. *Si has bebido agua del Colorado, una gran historia corre por tus venas.* If you have drunk of the Colorado River, a great history runs through your veins.

Everyone who lives along the Colorado River shares this history— Americans and Mexicans and Natives. The same water flows through them all. A part of this history that has not been well told is the role played by the river itself, not in colonizing the western deserts by irrigation, but in first opening them up. At a time when the boundaries of the West were still very fluid, the river was the highway, not just for seeds and beavers, but for human beings as well. Americans came up the river through Mexico to gain their initial access to the deserts in the southwestern part of the United States. In the process, the cottonwood-willow forests and the Cucapá Indians were indispensable to the Americans. Neither has received the credit they deserve in this history.

One of the most startling facts about Yuma, Arizona, perhaps the quintessential desert city in the United States, is that it was officially labeled a seaport city. No ship has ventured up the Colorado River to Yuma for the better part of a century. Steamboat traffic shut down along the river back in 1916. Yet Yuma is a seaport and steamboats opened up the desert—another of the paradoxes and ironies of the Southwest.

The book *Steamboats on the Colorado River 1852–1916*, by Richard E. Lingenfelter, tells the story of the riverboats best. To understand the role of the Natives in this venture of the opening of the desert—their role is almost always overlooked or given insignificant attention—I recommend Anita Williams' book, *The Cocopah People*. The related story of the cottonwood trees in opening up the desert has not yet been told.

Settlers of European descent largely eschewed this desert region of the continent. Only with the Gold Rush to California in 1849 did the desert begin to open up to exploration, and then to settlement. According to William deBuys, in his striking account of this horrifying route in *Salt Dreams: Land and Water in Low-Down California*, "thousands of over-equipped and ill-prepared whites . . . appeared at the Yuma Crossing in 1849" (deBuys 1999, 34). Beyond Yuma and the Colorado River lay the worst part of the desert journey on the route to "El Dorado," one-hundred miles of the horrifying desert full of the abandoned detritus of earlier "argonauts"—the settlers pictured themselves as being on a mythic journey akin to that of Jason and his argonauts in search of the golden fleece. The air was fetid with the smell of dead mules and misery. John W. Audubon, the bird artist's son, passed through this landscape. The journey broke him physically and spiritually. He later looked back upon his journey as an excursion into desolation.

It was to protect these fevered travelers that Yuma was established along the river. Only a year before the California Gold Rush began, huge portions of lands in the North American West had been wrested from the Mexicans by war and formalized in the 1848 Treaty of Guadalupe Hidalgo. Now Americans were claiming the lands in their growing empire. Yuma became one of the crucial, but most godforsaken, outposts in that empire, founded to protect the men and women of the Gold Rush. For soldiers, it became famous as the worst posting in the United States. One newcomer described the fierce desert as "hell with the fires put out" (quoted in Lingenfelter, 1978, 34).

The fort at Yuma was established to protect the ferry business that had sprung up at the river crossing. In 1849, a man named Doctor Lincoln began the service for the immigrants to California. He sold out to a band of American renegades and cutthroats in the John Glanton gang. Glanton and his men were scalp hunters from Texas. They ran a terrifying business of extortion, exorbitant prices, rape of Mexican women passengers, and attacks on the local Quechan or Yuma Indians. The Indians set up their own rival ferry crossing.

Glanton's gang broke up the Indian ferry, smashing the boat and drowning one of the men. In 1850 the Indians rose up against the Glanton gang, killing eleven of them, including Glanton himself. When three survivors in the gang straggled across the desert to San Diego, they reported a "massacre" of whites by

Indians and demanded retaliation. The governor of California dispatched Gen. J. C. Morehead to exact punishment on the Natives. He assembled a force rounded up by the county sheriff, and headed into the desert.

While Morehead and his men laid waste to Quechan lands, the military authorities sent Maj. Samuel P. Heintzelman to the Yuma crossing to get control of the lands and pacify the region. Heintzelman subdued the Quechan by crop destruction and starvation, and then established a fort on the west side of the Colorado River. He laid out a military settlement that included the ferry crossing and, in March 1851, established Fort Yuma. The settlement completely encompassed the ferry crossing on the river. When the owners who had taken over after the Galton gang protested, Heintzelman told them they should just sell out to him.

Heintzelman quickly realized that the biggest problem facing this lonely and hateful outpost was not Indians, but logistics. Living conditions were abominable—extreme heat and desolation. Hauling supplies two-hundred miles across the desert by mule train meant that the men lived on starvation rations. At one point, even camels were tried in the American desert. They failed, though, when they balked at crossing water.

These difficulties in supplying the fort for the Gold Rush argonauts led directly to the establishment of vigorous steamboat traffic up the river through the delta from the Gulf. It was this traffic that opened up the delta and made the settlement at Yuma viable. The idea of supplying Yuma from the sea and the river was also based on bad maps and mistaken geography. Maps of the delta at the time were miserable—few whites had traveled in the region since 1781, when the Quechan uprising had driven the Spanish missionaries out of the Yuma area. One British naval lieutenant, Robert Hardy, had entered the Colorado River in 1826 while searching for pearling grounds in the Gulf of California. He traveled about twenty-five miles up the river. When he arrived at the confluence with another river, he thought he had reached the Gila River. He had only made it to another arm of the Colorado River in the delta, subsequently called Hardy's Colorado or Río Hardy. But he did not know that.

Nor did Major Heintzelman. He thought Fort Yuma was an easy twenty-five miles from the Gulf of California. Running supplies by boat up such a short stretch would be neither difficult nor expensive—much less so than the $500 a ton it cost to freight supplies on mules across two-hundred miles of desert.

So a Lt. George Horatio Derby was dispatched in a schooner named *Invincible*—a hopeful and ultimately inaccurate name—to haul a load of ten thousand rations up the river to Yuma. He entered the mouth of the Colorado River on Christmas Eve 1850, and would soon be disabused of the optimistic faith in an easy passage upriver. Derby covered the expected distance and, of course, found no Fort Yuma, quickly realizing Hardy's error. He fired his guns in case the fort might be nearby. Then he dispatched Cucapá Indians upriver to find the fort—which was another 120 circuitous river miles away.

Meanwhile, Derby sat in the small schooner near the mouth of the river. It turned out that Heintzelman ran into Derby near the mouth of the river. The commander of the fort had set out south on his own reconnaissance mission. Neither of the soldiers was concerned with natural history. Nevertheless, Heintzelman notes in a letter on this trip that "the river bottom is several miles wide and covered with willow, cottonwood, and mesquite . . ." (Sykes 1937, 19).

The question that obsessed both Heintzelman and Derby was navigability. Derby and his schooner sat for some time at the mouth of the river, trying to figure out a way of getting upriver. Finally, his boat was badly bashed about by the huge bore, or tidal flood, at the mouth of the Colorado. One of the famous phenomena of the pre-dam river was that the large tides of the Gulf met the powerful currents of the river and formed a wall of water, as much as ten feet high, that worked its way backward up the river. The *burro* of the Colorado, the bore, would capsize thirty-six-ton steamboats.

Derby dumped his supplies on the Sonoran side of the river mouth and decamped. Heintzelman sent men back down overland to pick up the supplies, but this arrangement was worse than the route from San Diego through the desert. Besides, it was a "palpable violation" of the Treaty of Guadalupe Hidalgo to use Mexican land as a kind of ship port. One of the key provisions of the treaty guaranteed international navigation, but hauling goods through Mexican land was another matter. Supplies in Fort Yuma grew so wretched that Heintzelman was forced to abandon the fort.

The answer to the supply problem came finally with steamboats. They offered a technological solution, setting a pattern for all subsequent white relations with the desert. On December 3, 1852, the steamboat *Uncle Sam* completed a difficult, 120-mile journey from the mouth of the Colorado upriver to the Camp Yuma landing. After about two weeks of churning up the perverse currents and negotiating the shifting bars of the river, its deck heaped with some thirty-two tons of freight, the *Uncle Sam* became the first boat of the modern era to succeed in ascending the delta river. Along the way, it had even been badly shaken by an earthquake.

Two Cucapá Indian leaders from the lower delta were on board, helping to navigate the labyrinth of channels that are so easy to get lost in (Williams 1974, 37). For the next twenty-five years, an active river trade from the Gulf to Yuma opened up the delta and the desert Southwest. Because they knew the lower river and its mazes so well, the Cucapá people frequently served as river pilots.

To run the steamboats, huge quantities of firewood were required. The Cucapá served the river trade in another capacity as well, as *leñeros*, or woodcutters. During the heyday of the steamboats on the river, five separate woodyards were established along the river, each located roughly a day's journey, or thirty miles, apart. They were the only settlements below the Yuma ferry and the wood was completely supplied by the Cucapá workers. According to Lingenfelter:

> The first yard above the mouth of the river was ominously known as Port Famine; above that was Gridiron, then Ogden's Landing . . . and finally old John Pedrick's, just above the boundary line—the first landing in the United States. Though most of the yards were on Mexican soil, the owners were all Yankees. They hired Cocopahs [an alternative spelling for the tribe] to cut and haul wood to the river and were said to make as much as $5000 a year from the business. (Lingenfelter 1978, 12)

These woodcutters are compelling evidence of the abundance of the forests in the delta.

The *Uncle Sam* proved that the delta was navigable. It remained to be determined how far up the river was the head of navigation. The job of exploring the river by steamboat fell to Lt. Joseph Christmas Ives of the topographical engineers—under orders from the Secretary of War, Jefferson Davis. Ives' account is one of the most colorful and revealing pictures of the delta in the era before the twentieth century.

Ives may have been something of a dandy. Lingenfelter calls him a "foppish, self-aggrandizing fellow." He sported a broad moustache and slicked hair. In any event, he saw this assignment as the opportunity of a lifetime and his narrative is a delight to read compared to the usual army prose. He fancied himself in the long tradition of great western explorers heading into terra incognita. His little steamboat looked like a joke to Cucapá and other onlookers as he took off up the river. They called it the *"chiquito* boat." The artist on the expedition, Baron Möllhausen, called it a "water-bourne wheelbarrow." Not Ives. He gave it the name he wished for himself, *Explorer* (Lingenfelter, 1978, 16–17).

Ives' adventure was a bit comical, as his large dreams were regularly punctured by the reality of his experience. He followed the customary method of fitting his steamboat—hauling in pieces to the mouth of the delta and then assembling the boat there on the huge mudflats. A kind of seaport in the mud was set up there, called Robinson's Landing, where he could stay in the "Colorado Hotel"—really only a single-room shack on stilts.

His descriptions of the area are beautifully lyrical, and funny. He speaks of seeing "innumerable flocks of pelicans, curlews, plovers, and ducks of different varieties" while working at the mouth of the river, putting together his steamboat. "It was easy to shoot them," he says, and then falls flat on the facts of living in a mudflat: "but almost impossible to get them afterwards on account of the depth of the mud . . . ." And then there was also the sheet of water sometimes several miles wide that covered the entire mudflats (Ives 1861, 25).

*Early Mexican settlers in the delta called themselves "cachinilla," the name for the arrow-weed bushes once common in the desert and out of which they built their mud-and-stick houses.*

Despite the pressures of the place, Ives was not insensitive to its beauty at all, especially its light. It evoked poetry from him. "At sunrise the atmosphere is singularly pellucid," he writes, "and every point of the surface of the water and the land sparkles with light . . . and a dazzling azure glare" (Ives 1861, 29–30).

He goes on to describe how light seems to be one of the great realities along the delta river. At certain times of the day, it can seem the only reality:

> As the sun mounts higher, and the light becomes more intense, these mountains [Sierra Cucapá in the western distance] grow indistinct, and are gradually lost in a bright mist of grayish blue that seems to blend earth and sky. The nearer mountains, the water, and the flats, all partake of the same blue cast, and throughout the day are invested with a dazzling azure glare. (Ives 1861, 29–30)

Ives had all manner of trouble ascending the river. He and his crew only barely made it off the mudflats before a very high tide came to sweep everything away. They got stuck on sandbars in the shifting river. They lost their way. Finally, Ives abandoned the *Explorer* when it foundered hopelessly on one bar. He hiked overland in quest of Yuma, a perilous journey since he had no idea where he was. When he arrived at Yuma, he could hardly bear this outpost in oblivion. He called it the "Botany Bay of military stations," referring to the criminal outpost of Sydney, Australia.

Ives managed to ascend the river almost five hundred miles from the mouth of the river, cracking his hull finally on a rock at the entrance to Black Canyon. He christened it "Explorer Rock" and pronounced it the practical head of navigation.

A brisk business continued on the delta until 1877, when the Southern Pacific Railroad finished a route through the Colorado Desert and drove the steamboats in the delta out of business. A couple of boats continued to work the river north of Yuma, and occasionally made trips back down into the delta for firewood from the abundant cottonwoods and willows. But really, the delta sank back into untraveled obscurity for another two decades, which is how it was when Godfrey Sykes arrived and began his long, adventurous love affair with the delta.

One of the truly compelling photographs in the collection at the Museo Universitario in Mexicali shows two Cucapá men standing beside a stack of wood cut to fuel the steamboats on the Colorado River. Behind them and the wood, leaning against the bright cottonwood, highlighted against the depths of dark forest beyond, are about ten white men. It must be a photograph taken at one of the five woodyards along the river in the delta. The picture is particularly remarkable among all of those in the collection for what it adds to the story of the river and the delta. The story of the conquest of the arid West is typically one of heroic achievement and technology—the story, say, of Lieutenant Ives in his ascent of the river. Yuma and the settling of the desert was made possible by great explorers and their powerful steamboats, according to this tale. But this photograph reverses the customary iconography of conquest. The white men are in the background. The photo foregrounds for us the usually forgotten figures in the opening up of the West—two Native men who supplied the boats with stacks of Colorado cottonwood, and the wood from the forests, which fired the engines of the steamers.

The sun is only a glowing hint in the cobalt arc along the sandy horizon. The morning is still cool as we dodge the willow branches. Osvel Hinojosa and Jaqueline García jump off a small ledge of the riverbank and walk into the dried-up riverbed. The river no longer flows here. They are standing in rippled sand. They have also just crossed the border between the United States and Mexico.

In the gray-blue shadows of dawn, they spot a single track of human footprints through the otherwise untracked sand. The prints are heading the opposite way—into the United States. In a clump of small willows in the middle of the old riverbed, Osvel and Jaqueline find an abandoned tennis shoe and an old sleeping bag. An illegal immigrant had camped here last night, and then made a dash for a new life in the United States.

We are just south of the Morelos Dam, the only Mexican-built dam on the Colorado River. The dam was required by the U.S.–Mexico 1944 Treaty on waters because this location—just west of Yuma, Arizona and the Yuma Desalting Plant—is where most of the treaty-mandated deliveries of water to Mexico are made. South of the dam, twenty-four miles of the Colorado River define the border between the United States and Mexico. The line is called the "limitrophe" and separates Mexico and Arizona. When the Gadsden Treaty adjusted the boundary between the two countries in 1853, this stretch of the river became the international boundary.

There are no fences here. There's almost no river. Frequently the river boundary no longer flows here. Between the lack of flows over many years, and the intriguing fact that when it does flow through here the river has changed course, the limitrophe offers a case study in the Tao of boundaries. The river was the boundary, but disappears some years and over time has changed course. It is a dry river, perhaps, but it reveals a fluid boundary.

For people charged with guarding the boundary—Border Patrol and the International Boundary and Water Commission—the border problems are practical issues. By treaty, this stretch of boundary is supposed to be redefined every ten years. It has been twenty-nine years since the last redefinition. What they want to do is solve the problem once and for all by digging a channel for the river and lining it with concrete. Both river and boundary can then be brought under more complete control.

The porous border is a metaphor for the arbitrary nature of boundaries. It suggests to me that the two sides of the border have conjoined destinies in this single delta watershed. For certain migratory birds, the limitrophe is not a boundary between countries, but prime habitat in their migration route. That is why we are here this morning. We did not come to find Mexican illegals crossing into the United States. We came to find one of the prime endangered species in the region—the southwestern willow flycatcher.

Jaqueline has asked Osvel to help her in this limitrophe region with a June survey for migratory southwestern willow flycatchers. Though her Ph.D. is on chemical contamination in the delta, she has been hired on contract work by the Cocapah Indian Tribe in Somerton, Arizona—the United States branch of the delta tribe of Indians. Their interests in knowing and preserving the delta extend beyond the boundaries of their particular studies. That is why they are doing this survey.

As we hike along the riverbed in the early light, Jaqueline and Osvel both play tapes of the southwestern willow flycatcher's call. A small bird with a drab olive green plumage, the southwestern willow flycatcher is one of the premier endangered species along the Colorado and other rivers in the desert Southwest. It is smaller than a robin, sits upright on exposed branches, and makes short darting flights into the air, where it snatches bugs on the wing. Hence it's a "flycatcher," a large family of insect-eating birds.

Though nondescript and often a challenge to identify, flycatchers, with their alert eyes scanning the air and their darting aerial sallies, never fail to raise the heart. As important, the close association with wetland habitats—they are always found in woodlands closely associated with water—has made it the endangered poster child for disappearing wetlands in Arizona and the Southwest. There are about three hundred to five hundred nesting pairs of the southwestern willow flycatcher in Arizona, and nearly every one of them is known and mapped by biologists—so precious have the birds of this species become. Most of the nests are in the riparian habitat along the San Pedro River. The birds are not yet known to nest in Mexico, but they definitely migrate through the delta, as this census is helping to demonstrate.

Most *Empidonax* flycatchers are hard to distinguish from each other. The willow flycatcher is especially difficult to tell from the Trail's flycatcher—the one we caught in the cottonwoods in Mexico. So once you know it is either one or the other, the best way to identify the southwestern willow flycatcher conclusively is through its call, an explosive *fitz-bew*. Usually, it gives this call from an exposed branch on which it perches, looking for insects.

That is the call Jaqueline is playing in the riverbed. It is early, but a bird answers from a nearby willow tree. We search and find it, a dark shape on a dead branch of the tree. As the surveys would show, this 170-acre patch of willows along the river is crucial habitat. On this single day, Jaqueline and Osvel and I would count fourteen separate flycatchers. They were all on migration through this riparian corridor. Another census a week later in June would turn up twelve more southwestern willow flycatchers. But a census in late June turned up none. The birds are not nesting in the area, but passing through. This area is prime habitat in their migration route.

This endangered flycatcher demonstrates emphatically the value of the riparian corridor along the river. This is one of the areas inventoried by Carlos Valdés. We are looking for the birds on the U.S. side of the river, but a corresponding habitat exists across the boundary. Technically, we are not looking for birds in the United States. All of the habitat is within the third nation in this complex of territories and water rights along the border. The third nation is the Cocopah Indian Tribe.

In 1917, President Woodrow Wilson awarded reservation lands to the Cocopah Indians in Arizona. The lands are adjacent to the Colorado River in the limitrophe. The Cocopah have one large plot of land near Somerton and another up closer to the Morelos Dam and the Yuma Desalting Plant. These Indians are not only another stakeholder in the complex jockeying for water rights. Like another twenty-six tribes of Indians on the Colorado River, the U.S.

Cocopah have rights guaranteed by complicated legal decisions, most notably the decree by the U.S. Supreme Court in *Arizona v. California.* Yet the environmental officer of the tribe, John Swenson, has contracted Jaqueline to conduct this survey as part of the tribe's biological inventory of the reservation lands. The tribe is considering using some of their water to create wetland and riparian habitat for conservation.

The Cocopah in the United States are becoming important players in the water issues of the Colorado River. The question facing water managers increasingly is how to accommodate legitimate "water users" like Native Americans, who have long-established legal claims, as well as endangered species, which currently have no such established legal rights to water. Along the border, the issue of taking into account the environment has two dimensions: one, within the United States; two, transboundary environmental issues of the kind posed by the delta. The first has been its own challenge. The second is more difficult. Creatures like the endangered southwestern willow flycatcher bring the importance of considering the creature's entire life cycle into sharp focus. What good does it do to try to protect the species in the United States, and then not protect migratory routes through Mexico?

All of this is part of the larger complex of social and political issues at the border—a complicated ecology of humans and creatures along the border, legal and illegal migrations.

Lives in motion, lives on the border. This whole border region has been changing rapidly over the last few decades. Both the political and the natural environment have changed. Hispanics, of course, are becoming a defining element in border life in the United States, as is widely documented. The census of 2000 shows that Spanish language use in the United States has grown by 50 percent. We are not as separate and distinct from Mexico as we used to try to be. Perhaps we need a more enlightened relation with Mexico that is at once social and environmental.

Perhaps we need a new conception of the borderlands—not simply as a place of exclusion and control, but also of contact and connection. The model might be based less on the notion of "us versus them," and more on a notion of "we." We are connected across the boundary by culture and history as well as by ecology. I am reminded of the anthropologist James Clifford, who, in his book *Routes: Travel and Translation in the Twentieth Century,* tries to construct a new way to talk about the interchanges between cultures. Cultures have become fluid in this globalized age. He is interested not in borders but in "a borderland, a zone of contacts—blocked and permitted, policed and transgressive." He explains how a new way of thinking about the boundary area has emerged in recent years, largely the result of the "Latinization" of the borderlands:

We need to conjure with new localizations, such as the "border." A specific place of hybridity and struggle, policing and transgression, the U.S./Mexico frontier has recently attained "theoretical" status, thanks to the work of Chicano writers, activists, and scholars: Americo Paredes, Renato Rosaldo, Teresa McKenna, José David Saldívar, Gloria Anzaldúa, Guillermo Gómez-Peña, Emily Hicks, and the Border Arts Project of San Diego/Tijuana. The border experience is made to produce powerful political visions: a subversion of binarisms, the projection of a "multicultural public sphere (versus hegemonic pluralism)" (Flores and Yudice, 1990). How translatable is this place/metaphor of crossing (Clifford 1997, 37)?

What he means is that a new vision is being created out of the experience of life in these borderlands. With it, new metaphors are emerging—and metaphors are crucial for determining how a culture conceives of itself, understands its experience, and establishes relationships. The old metaphors of us and them—separation, displacement, and control (Clifford's "binarism" and "hegemony")—are giving way. In their place, a new metaphor of a "zone of contacts" is emerging—migrations, contact, and interaction, describing a more inclusive sense of relationship.

Something is missing from Clifford's anthropological "we," though. The southwestern willow flycatcher passing through the regenerated border corridors is every bit as much a part of the border crossings as the people traveling back and forth—the legals and the illegals, the privileged and the persecuted. In the emerging new maps of these border crossings, these two groups parallel and mirror each other in this "zone of contacts"—the animal and natural other, as well as "the others" who are people.

The task is to integrate an environmental concern that transcends international boundaries with changing cultural values. The trick is to find ways to give the southwestern willow flycatcher legal recognition in the United States, not just when it is nesting here in Arizona, but also when it is migrating through the willows and cottonwoods in the Mexican delta.

Later that same evening, after leaving Jaqueline and Osvel, I returned to the limitrophe to do more bird-watching. The willows produced a lot of birds: Costa's hummingbirds, hooded orioles, rough-winged swallows.

Road runners.

Not just the avian kind, either. I was driving along the levee road beside the Yuma Desalting Plant. The river is very close here, not visible put a presence felt in the stretch of reeds and willows to the west. Morelos Dam is hidden in a thicket. Out of the thicket I suddenly saw two young men race into the open. They followed dirt paths through a small field of alfalfa and ran right below me on the dirt road that follows the canal carrying irrigation water. In the background was a trailer park and a golf course, both of which are enterprises owned by the Cocopah Indians. They rent the trailers to northern "snowbirds" who come to Yuma to golf through the winter. In the distance, across the Colorado, were the Algodones Sand Dunes.

The two men, certainly in their early twenties, ran right past me. One wore a Pittsburgh Steelers football jacket, black and gold. The other carried his red baseball cap in his hand. They looked already Americanized. But I realized they were Mexicans, illegal immigrants making their move across the border. I expected to see the Border Patrol come screaming out of the willows, huge clouds of dust flying. Despite all the high-tech surveillance on the border, nobody came. I remembered a joke Osvel told me. I have heard it often as I travel in Mexico. Mexicans they say are winning back their territories in the United States, without having to fire a shot.

Marginal lives along the border —animal and human—both suggest there must be a better way for us all to live together.

As the Mexicans ran below me, I yelled and waved. I am not sure why. I wondered if they had family or friends they would join once they got beyond immediate danger here. One of them, the kid with the red cap in his hand, looked back at me. He waved back, his red cap vivid against the green agricultural fields. He did not break stride. At a full run, both kids disappeared around the side of the levee.

*The limitrophe is the section of the river forming the border between Arizona and Mexico and is one of the best wild habitats in the delta. Footprints through the sand in the predawn tell the tale of a human immigrant that crossed over last night.*

143

# EIGHT

# "We Are Not Yet Dead Still"

ENDANGERED PEOPLE: Cucapá Indians
SPECIES: Coyote
PLACE: Río Hardy and Laguna Salada

NOT EVERYONE WAS HAPPY for the massive floods that swept out of the dams and over the delta following the El Niño years in the 1980s. Among the people who suffered most were the Mexican Cucapá, the same ones who helped the Americans open up the river.

"We are the people of the river," says Monica Gonzalez Portillo. She is Cucapá and they still refer to themselves that way, *la gente del río*. They were the last of the tribes of indigenous people living along the river. The Colorado River shaped their lives. They lived seminomadic lives according to the river's stages through the year. They farmed in the muddy waters of the delta. They lived on the fish the river provided.

That way of life is gone. Monica and her tribe live now in the village of El Mayor at the base of the granite mountains named for their people, Sierra Cucapá, about thirty-five miles south of Mexicali. Their village is dusty and bleak and sad. It looks like they are living in a town plopped down in the middle of the mountain's scree. Jagged rocks cover roads in town. Shanty houses warp and fall apart in the sun, unrelieved by trees or shade. The river is nowhere in sight. The river people no longer have water. Now they have to buy it in plastic jugs from trucks that come through their town. Pick-ups disintegrate in the yards, cannibalized for parts, rusting from the salts in the soils.

Early Spanish explorers reported as many as twenty thousand Cucapá natives living at the end of the Colorado River. That number is almost surely

*◄ The Sierra Cucapá are mountains sacred to the Cucapá people in Mexico. They rise above the dammed portion of the Río Hardy on the west side of the delta.*

*▲ When the desert blooms, the Cucapá harvest the flowers of the biznaga or fish-hook cactus for food.*

145

inflated. More reliable is the estimate by the Spanish missionary, Father Garces in the late 1700s, who estimated a population of about eight thousand Cucapá. These people thrived along the river. Now Mexican Cucapá number only about three hundred people. Seventy of them live in about forty families in this sad town.

*"Tierras malas,"* Monica says, looking around the village. Bad lands. You might think of it like some rural desert ghetto. Monica is in her early thirties, a gentle woman with a lovely round face and deep beautiful eyes. She has just given birth to her second child, a young daughter, Alías Madasdana. Her husband Elias is from the Kiliwa tribe near Ensenada. The baby's name means "half moon" in his native language. Monica still radiates the deep satisfaction of having given birth.

It is easy to get carried away by the external signs of desolation in the town. And if you focus on that, you will miss the town's real beauty—the people.

I fight back a rising sense of anger as Monica tells me the story of how her people wound up in El Mayor. As she speaks, I realize that there is a feisty spirit in her. There is a defiance in the people that is belied by the external appearances of the village. They are not after sympathy. They want recognition of their human rights. They want their lives back. They want their river back.

Monica's face is sweet and happy, her voice soft. But they both mask a steely resolve as she describes her tribe's condition. "When the settlers came and took the river," she says, "no one so much as asked our opinion."

Even after the dams went in upriver, the Cucapá continued to live along Río Hardy, an arm of the Colorado River about sixteen miles long in the delta. It was part of their traditional lands. They lived in loosely scattered houses along the riverside. Then the dams filled, El Niño hit, and the American dams opened their penstocks. The floods rushed into the delta.

Monica was ten years old. She says she can remember the floods. The waters poured over the riverbanks. They lost their houses. They put everything they owned in boats, she says. Then the government made them move to this village on higher ground in the scree fields at the base of the mountains.

"I cried when we had to move," Monica remembers. For months they lived in tents. Then the government helped them get materials to build some houses. They only recently got electricity in the village, but most can't afford it. Unemployment is very high. There are no medical services. Disease is a growing problem. Most devastating, they have no rights to water. Unlike American Indians, only one tribe of Mexican Natives signed a treaty with the Mexican government. Monica says the prejudice against indigenous peoples in Mexico is intense. This tribe, the "people of the river," has no recognized treaty rights to land or water in the delta.

After decades of struggle, the Mexican government finally recognized the Cucapás' rights to land. The tribe was awarded 143,000 hectares of land,

357,500 acres. It seems like a lot. But none of it can be farmed. It is all in the part of the delta that is either desert or has been turned to desert by salts and fertilizers and other poisons. Even on their land, the Cucapá do not have the right to fish in the waters. The government owns all the waters that flow to the sea in Mexico. They allow everyone to fish on waters that might come in flood years to Cucapá lands.

"Everything we have we have to fight for," Monica says, nursing her baby. "The government protects species of animals so they don't go extinct. But us, the Indians, we are also in danger of extinction, and nobody protects us."

The story of the Cucapá in Mexico is one of the saddest stories of conquest. Displaced and dispossessed, they were not conquered by an army. They were famous for being friendly to the settlers who moved in. The Cucapá worked for them and helped them, including the Americans who pioneered the delta in steamboats. What nearly destroyed the Cucapá was the same thing that forced them to live in this poor little town of El Mayor—the dams on the river. It was a subtle form of conquest. The historian Alfred Crosby speaks of "ecological imperialism," in his excellent book by the same title. He describes the way native plants and animals, as well as Native peoples, were subjugated indirectly—by introduced creatures and by, say, taking their water. His concept would apply to what has happened to the Cucapá.

The Cucapá are the most striking example of environmental injustice I have seen. And they live within a three-hour drive of San Diego. They are extremely friendly and gentle. Once they flourished along the Colorado River, but now have been completely forgotten by the civilization on both sides of the border that destroyed their old way of life.

"You know that the river is life. You also know that when there is no water, there is nothing." *Tu sabes que el río es la vida. Tu sabes también cuando no hay agua, no hay nada.*

Monica looks at her baby, the image of the future of the tribe. For better or for worse, the history of the Cucapá has been tied to the history of the river, their fates joined. Now they are fighting to get more water for the river and they are fighting for their rights to the river. It is that fight I have come to learn about. I am going to join Monica's father, Don Onesimo Gonzalez Sainz, who is the *Jefe Tradicional* or Tribal Chief of the Cucapá. He is going to take me out for the opening day of fishing on the river, where the Cucapá are fighting to get their rights back.

As the day begins, I have no idea how much more he will also reveal to me about nature and the human spirit in the delta.

Don Onesimo is a cultural treasure to the Cucapá. He was born in 1934. Hoover Dam came on line in 1935. Don Onesimo's life almost precisely spans the period of the delta after the dams. He remembers part of their earlier life. He has suffered through the tribe's plunge toward extinction. He is a living, oral history of the Cucapá people through the twentieth century.

As he puts it, he has spent a lifetime fighting everyone for his people's rights. He still speaks his native Cucapá and he tells me the native names for everything. "The river's almost finished," he says. "Everything's changed."

He is a small man now, with fine features. There is a diffidence and a gentleness in his manner. And a friendliness and generosity that are disarming. He is eager to talk about his culture. He wants to get the word out. It is one of his missions in life now, along with passing the tribe's culture to its new generations. By his testimony and work, he hopes to preserve his endangered tribe and its culture.

I visit him at his home, where he invites me to stay for several days. He lives a few miles from El Mayor along Río Hardy, in an enclave that is the broken-down remnant of better times. The name of the little place where he lives is *Campo Flores*, Camp of Flowers. The name is almost savagely ironic. Once this place was a cluster of vacation homes for Americans who came down from San Diego and Los Angeles, when Río Hardy was still vibrant.

Even as Lake Powell was filling behind Glen Canyon Dam during the 1960s and 1970s, Río Hardy had stayed green and alive. At the confluence of the Hardy and the Colorado Rivers, downriver from here, a natural river bar had blocked off the Hardy. The backed-up waters had created a huge wetland complex on this western side of the delta. That's why the Cucapá were living down by the Hardy. Campo Flores grew up on the river during that time, remaining an oasis on the delta even as the rest of the delta withered and dried. Many of these camps opened up along Río Hardy, catering to Americans who wanted to hunt the abundant ducks and geese in the wetlands or fish the river.

*Don Onésimo González is Jefe Tradicional, chief of the Cucapá in Mexico, preserving the language and stories of the original "People of the River" and fighting for their survival. He stands before the Río Hardy.*

Then came the floods of the early 1980s. Not only did the floods drive the Cucapá from their homes, they also wiped out the bar at the mouth of the Hardy. The wetlands drained as the Hardy shrank from two hundred feet across to about twenty-five feet. The whole area turned to salt. Some landowners convinced the Mexican government to build flood levees up higher on the Hardy. Those owners now have a nice little lake, one of the few sweet spots still in this part of the delta. But that's upriver. El Mayor and Campo Flores are below the levees. They were forgotten.

"Above the levees, they eat," says Don Onesimo. "Below the levees, no."

Now Campo Flores is eerie and empty. A few Americans continue to visit, dove hunters mainly. But not many. The houses are sad and neglected. Don Onesimo lives here virtually alone in a small house beside the river. He rents it for about $250 per year. It's painted in a kind of aqua blue that is itself a relic color from the 1960s. The house is falling apart. The screens on the windows are ripped up. No electricity. Running water comes straight from the Hardy. That is a scary thought in itself.

As the Hardy shrank, it turned into a frightening brew of poisons. Flowing below Don Onesimo's house, it is the color of automobile antifreeze. Every drop of the 200,000 acre feet per year that flows through the Hardy is now agricultural runoff water or wastewater from the geothermal plant up by Mexicali. There are seventeen agricultural drains in the Mexicali Valley. Literally tens of thousands of tons of fertilizers, and about 100,000 gallons of insecticides, are poured onto lands in the lower Colorado, according to Mexican authorities. Much of this chemical waste finds its way into the Hardy.

High levels of selenium, boron, and arsenic have been found in the birds from the Río Hardy. Dissolved selenium in these waters is up to fourteen times higher that the U.S. Environmental Protection Agency criterion for protection of freshwater life, according to the studies by Jaqueline García. Below the confluence of the Hardy and the Colorado Rivers, the salinity in the water doubles. Cottonwoods can't grow below the Hardy any more.

"Los blancos han envenado el río," Don Onesimo says angrily. The whites have poisoned our river, referring to both Americans and Mexicans. "Aguas negras," he says, a vivid description of polluted water: dirty waters.

"Some of our people have white grains appearing on their skin lately," he tells me.

There are still a few pockets of wetlands in the area. It is not completely dead here, as Osvel Hinojosa and Carlos Valdés discovered, searching the area for patches of tule fed by springs. But mostly the salt cedar has taken over. This species of tree, tenacious as a weed, has thrived in salted-out land where few other plants can grow. Carlos estimates that it covers about fifteen thousand acres along Río Hardy. Introduced in the 1880s in Arizona, it swept quickly into the delta, sucking up huge amounts of water—a very thirsty plant. It is almost impossible to eradicate salt cedar. The easiest way to get rid of it, though not entirely effective, is to provide large amounts of freshwater.

That evening we sit around the Coleman lantern in Don Onesimo's house. We talk. Or really, Don Onesimo talks and I listen, enthralled. The entire camp is vacant except for us around the lantern. The house is full of shadows. Tomorrow is the opening of the fishing season for Gulf corvina, a kind of sea bass. Opening day is one of the major events in the life of the Cucapá, since the fishing season is the principle source of their current income. But they have been told that they can't fish in their traditional fishing grounds. So the Cucapá are fighting for their rights and fishing in defiance of the Mexican government.

"We're fighting the invaders still," says Don Onesimo.

His tone is defiant. Despite the sense of having lost nearly everything, despite the weariness in his voice, he has not given up.

"We're almost finished," he says. "What is done to the river is done to us. No one recognizes the rights of the Indians in Mexico.

"But we're alive. We are not yet dead still," he asserts. "We are here." This is a literal translation of his Spanish. *Pero somos vivos. Todavía no somos muertos ya. Aquí estamos.*

We both look around at the shadows. Tomorrow we have to get up early for the fishing. Silently we go to bed for the night, while outside the poisoned Río Hardy slides slowly past the ghostly houses.

We are jolting through the vast tidal plain of the delta. The old pick-up seems to lack springs and we are flung all over the bed of the truck as we careen across the rutted mud of the road. We have just left Indiviso, a town far to the south in the delta. It is the last town in this part of the delta, a town as close to the way the original Mexican settlers in the delta lived as still survives. Houses are still made of *cachinilla*, or arrow-weed branches, and mud. Several Cucapá also live here and we are going fishing with them. We are on our way to one of the traditional Cucapá fishing camps on the river, located beside the river in a vast empty tidal plain, which the sea floods from time to time.

Don Onesimo is in the cab of the truck. I am in the back with Pedro Bueno Bueno, a member of the Cucupá in Indiviso who has one of the most wonderful names I have ever heard—Pedro Good Good. A recent spring rain has turned the ride to the fishing camp into a jaw-rattling experience. The deep tracks across the mud are a dangerous quagmire, easy to sink into and get permanently stuck in if you slow down. It is easy to get lost out here, too, if you don't know the way through the maze of ruts. So we race across the flats at about sixty miles per hour, taking a pounding.

But Pedro Bueno Bueno is in heaven. His personality is as exuberant as his name. He must be at least sixty-five years old, maybe even seventy. His hair is completely white, his legs a bit shaky. Yet he enjoys the ride for all it's worth. He stands up in the back of the pick-up bed, looking a bit unsteady to me. He grabs the tailgate, then takes his white cowboy hat off and waves it in the air as we hurtle through the bright sun, as if he were riding a prize bronco.

It is the opening of the corvina fishing season and he is happy to be getting out on the river. To the Cucapá, the river is not only the source of life for them. It is sacred. It is the meaning to their lives. Don Onesimo explained this to me as we were preparing to get in the truck back in Indiviso.

*"El río es sagrado porque nos da alimento,"* he told me. The river is sacred because it feeds us. He tells me that fishing was the principle activity of the Cucapá when the river still flowed. They would move their camps down to the river in the spring and summer to catch fish, living there for months at a time.

Then the Mexican government forbade them to fish in their accustomed places on the river. When the dams cut off the flows of water in the river, most fisheries in the Upper Gulf went into decline. One of the great fishes of the Upper Gulf, the totoaba, nearly went extinct. Partly to protect the totoaba and other fish in the Gulf, the government created La Reserva de la Biósfera Alto Golfo y de California Delta del Río Colorado in 1993. As is customary in biosphere reserves, it was divided into two zones. One is called a buffer zone, where people can live and pursue subsistence activities like fishing. The other is the nuclear zone, where all activities except scientific research are prohibited—and that only by permit.

Fishing was banned in the mouth of the Colorado River and in the main channels. Mexican fishermen have been furious over the prohibition. The region is poor and fishing is a central source of income. Local residents have risen up in protest as the biosphere reserve has tried to enforce the bans. Don Onesimo describes the Mexican fishermen as *gente brava,* wild men. They have torn up their tickets, burned enforcement officers' trucks, and threatened worse violence.

Mostly what they do is ignore the prohibition.

At first the Cucapá were allowed to fish from their traditional camps, even though the two camps are inside the nuclear zone. Then they were suddenly told they could no longer fish there, according to Don Onesimo. He said the government told them that they were using modern ways of fishing—fiberglass *pangas,* or boats, and outboard motors. For Don Onesimo and Monica, it was just another sign of the prejudice that they have fought all their lives. For the struggling Cucapá, the prohibition against fishing was tantamount to a death sentence.

"If we don't fish, we die of hunger," says Don Onesimo.

They protested, demanding their right to fish in their traditional places, even in the biosphere reserve. They organized the villages of Cucapá and defied the fishing ban. Last year Monica organized the protest. They went out fishing on the river anyway. It was an issue that caught the attention of every newspaper in northern Baja California and Sonora.

"We're going to fish until the very end," Monica is quoted as saying in one newspaper story. "For us this is a road with no exit."

They were confronted on the river by agents from Procuraduría Federal de Protección al Ambiente (PROFEPA), the enforcement arm of Secretaría del Medio Ambiente, Recursos Naturales y Pesca (SEMARNAP), the ministry in Mexico that manages environmental and fisheries matters. (Recently, fisheries have been split off of this ministry, which is now strictly charged with environmental issues. The agency for the environment is now referred to as SEMARNAT, the final letters of which stand for "Naturales".) The agents refused to let them fish.

Don Onesimo says that the Cucapá were singled out. The Mexican fishermen were on the river, too. But the police did not stop them. According to Don Onesimo, officials only stopped the Cucapá. So the Cucapá, under Don Onesimo and Monica, made a bold move. They filed suit, a *denuncia*, against the Mexican government. The Cucapá have no legal treaty rights, as American Natives have. The suit claims, however, that under Article 4 of the Mexican Constitution, the government guarantees to protect and promote the lives of Mexico's indigenous people. The lawsuit claims that the Cucapá must have their right to fish in their traditional grounds to protect and promote their lives. It is a basic human rights issue for them. They have filed another suit in which they claim that they deserve to be given water rights to the Colorado as well, as part of these basic human rights. Defenders of Wildlife in the United States is actively helping the Cucapá in this suit. It was filed through one of Mexico's top young environmental lawyers, Claudio Torres Nachón, and his environmental organization, El Centro de Derecho Ambiental e Integración del Sur, A.C. (DASSUR).

Despite the long history of repression for Mexican Natives, Don Onesimo is hopeful that the new government under Vicente Fox will be more friendly to Natives. In the meantime, the Cucapá won a temporary and unaccustomed victory, at least. A judge issued an *amparo*, or

*Pedro Bueno Bueno watches from the bank at the traditional Cucapá fishing camp as two fishermen head out onto the Colorado River.*

injunction, allowing the Cucapá to fish in their traditional camps this year. That is why we are going fishing. That is why everyone, including Pedro Bueno Bueno, is so joyful about going to the river.

When we arrive at the fish camp, we encounter a melee of activity and enthusiasm. Whole families are on the riverbank, setting up tents, installing cleaning tables for the fish, launching boats, or *pangas*. The river here is at least one hundred yards wide, flowing through a vast chocolate-colored mud-flat. The river itself is muddy and dark—not red, though. It flows between deep banks of mud, rising and falling with the tide. The mud in the river is not from silt. Rather the mud crumbles from the banks and falls in big chunks into the slurried waters. We are literally watching the delta erode.

The fishermen launch their boats with practiced flare. I watch one boat being launched just as we arrive. A man climbs down to the river, which is about twenty feet below us though rising fast. Using a cut-up plastic milk carton, he flings river water up the bank, making it slick as ice. Other men back a pick-up up to the edge of the bank. There is a *panga* in the truck bed. Slowly they shove the *panga* out of the truck. It teeters on the edge of the riverbank, and then careens madly into the river with a muddy splash that makes the men cheer.

Many boats are out fishing already. People are streaming to the river in more cars and trucks every minute. More *pangas* go crashing into the water. It is as much a social scene as it is a work scene.

Everyone is gathered to catch the Gulf corvina (*Cynoscion othonopterus*), a fish endemic to the Gulf. What is remarkable about the great harvests of this fish lately is that no one is sure what this species is doing here. The other fisheries in the Gulf began to plunge in the 1960s and 1970s. But the Gulf corvina staged a startling recovery that no one can quite explain. By 1993 it had come back in sufficient numbers to open up this fishery. One theory is that its comeback is due to the influx of freshwater into the Gulf from the recent floods. Another is that shrimp boats no longer work the nuclear zone any more. So maybe the juvenile corvina in the nuclear zone of the biosphere reserve are protected, leading to larger fish stocks.

At any rate, the recovery has been a boon to the Cucapá as well as to the Mexican fishermen. Even though this is a Cucapá fishing camp inside the biosphere reserve, a lot of Mexicans are here fishing as well, violating the ban that is still in force for them.

But no one says anything. The mood is festive. We watch several men haul in some fish out in the river. Another returns to the bank and flings a couple of big corvina up to his wife. Though it is a species of sea bass, silvery and shiny, it reminds me of a salmon. The woman proudly holds it up for me to see.

This is why the river is sacred to Don Onesimo, because it gives them fish. He tells me that they have only eighteen *pangas*. With such a small number of boats, the Cucapá will not put pressure on this fishery here, he says, if they are allowed to continue fishing. The fleets out of El Golfo and San Felipe caught five thousand tons of Gulf corvina last year. He says what the Cucapá catch in their *pangas* here is *nada* in comparison. Besides, he says, it is their only real income in El Mayor.

The most significant sign of hope for the Cucapá came a year after this fishing trip. The Comisión Nacional de Los Derechos Humanos (the National Commission for Human Rights) visited the Cucapá to research the tribe's *"denunsia."* In spring 2002, the Commission formally concluded that the Cucapá's human rights have been violated. It issued recommendations to the Secretary of the Environment, Victor Lichpinger, demanding that Semarnat recognize the tribe's right to fish in its traditional places, as well as its right to health and economic development.

The Commission's recommendations are a huge victory for the Cucapá.

Whatever the final outcome of their lawsuits, Don Onesimo says they will keep on fishing here. Don Onesimo looks around at the fishing camp. He and Pedro Bueno Bueno shake hands. This scene is about fishing. It is also about their temporary victory.

These people will not go quietly from here. The place is too important to them both economically and spiritually. "They'll have to drag me out of here," Don Onesimo says in colorful Spanish that defines his mulish resolution, *"con patas delantas."* With my legs forward.

As I travel through the delta and stay with Don Onesimo, I become increasingly aware of the privilege I enjoy in being in his company, learning the delta through his eyes. In one sense, my relationship with Don Onesimo is much like those with the many others I have met in my travels through the delta. He is an expert and I am a learner. He teaches me about the problems the Cucapá face in their struggle for social justice.

But something much more than that touches me in being with him. Something about the sadness that surrounds his life moves me, though he is never self-pitying. The issues he is struggling with are not abstract or scientific issues. They are life-and-death issues for him and his people. He speaks always from the heart about what matters most deeply to him. His reason for living, he tells me, is to keep the Cucapá culture alive. His wisdom grows out of a lifelong battle against extinction—not as an abstraction for animals, but as a reality for his people.

What I want to learn from Don Onesimo is not information, but rather his way of life and this human relation to the delta. One that has nearly vanished.

One afternoon he says he wants to show me something in the desert. *"Mucho misterio en el desierto,"* he says. So we drive down a side road into the *bajadas* of the

*A cactus garden of flowering biznaga grow in what looks like an avalanche of granite boulders. For the Cucapá, the mountains are full of mystery.*

Sierra Cucapá. The bases of the mountains are rocky. A few paloverdes manage to survive in the boulders and the sand, and gnarled mesquites. Soon the road peters out in the sand and we get out and walk.

We cross a sandy wash and arrive at a large field of rounded granite boulders.

155

They look like they have tumbled out of the mountains in a landslide and then have lain here for centuries, arrested and eroding in the wind. Somehow, out of cracks among these boulders, a whole miniature forest of cactis has grown up. The plants almost look like they are growing directly out of the boulders. They are small cacti, the tallest are only as high as my waist. Don Onesimo called them *biznaga*, fishhook barrel cactus. Most of them are in bloom, each crowned by a ring of honey-yellow flowers.

"We eat the flowers," Don Onesimo explains. "Oh yes, yes," he goes on. "There's a lot of food in the desert. A lot of mystery." *Mucho misterio.*

In this small cactus garden in the desert, many of the flowers have already been harvested. Don Onesimo tells me that several Cocapah from Arizona came down last week to gather them. He says they will make a *guisante*, a sauce, out of them. It is traditional food.

The earlier life was very pretty, he says. *Muy bonita.*

He explains that when he was young, he and his people would still plant food on the floodplain of the river. They did not build irrigation canals. It was floodwater farming developed by tribes along the river. The river brought fertile sediments down each year in the spring, flooded the mud, and receded. In the wet soils left behind, the Cucapá planted crops. They grew melons, and beans, and squash. They grew several kinds of beans.

Gary Nabhan tells of one farming ritual of the Cucapá in his excellent book, *Gathering the Desert.* Old men would plant panic grass in the mud. They used a ritual ceremony, planting seeds along with a badger claw, blowing the seeds from their mouths. By the time Don Onesimo was born, this grass was gone, along with the ritual. But he remembers planting other crops near Indiviso.

One of the crops the people relied upon down in the intertidal mudflats was a kind of wild grass, or wild rice, called *pan gentil*, another plant endemic to the delta. The botanist Edward Palmer discovered it in 1889 on a visit to the tide flats. It once covered 33,000 acres at the very end of the river. *Pan gentil* is now found on about 2,250 acres. It is a remarkable grass perfectly adapted to life at the end of the river. All it requires is one flow of freshwater in the spring to germinate. The rest of the year it can survive on the salt water of the sea. The Cucapá harvested the grains of this plant as a staple in their diet.

When I mention this grass to Don Onesimo, he gives me another version of its importance to the Cucapá. Scientists, he explains, only speak about physical things, *lo material*, the material world. They don't explain *la cosa misteriosa*, the mysterious thing.

Then he tells me a story about *pan gentil*, the people's bread.

Before the people could live down by the river, he tells me, they had to live up here in these mountains, the Sierra Cucapá. This is one of the reasons the mountains are sacred to them; it is where they came from. Everything down below the mountains, the whole earth, was wet. There was mud everywhere and it was very soft. Even though the sea went down a bit, Don Onesimo says, it was still very wet and muddy all over.

So the people on the mountain sent the horned lizard, *cameleon*, down out of the mountain to the wet earth. *Cameleon* looked all over in the mud. He went far south toward the sea and the river's end. There he found this grass. It had spikes on it, *espicas*. *Cameleon* grabbed three of the spikes and put them on his head to show them to the people. When he came back to the mountain, *cameleon* explained to the people that the seed low in the mudflats was ripe. But the people saw that *cameleon* had mud on his feet, so they knew it was still too muddy to go down. Instead they sent down ants, who worked in the mud to dry it out.

The story tells why ants and horned lizards are connected to each other. More important, Don Onesimo says, the horned lizard showed the people they could eat the grass and live in the lowlands of the delta.

This is what Don Onesimo has brought me out to this cactus garden in the desert to show me—the sacred nature of the delta to his people. The Sierra Cucapá is one of the people's sacred places. They have a geography of sacred places in the delta—a landscape invisible to whites, for whom the delta was a wasteland to ignore or exploit. For Don Onesimo, it is not just the physical delta, which feeds the people, that is endangered. It is also this sacred landscape, which sustains the Cucapá culture, that is endangered.

Don Onesimo explains to me that it is better not to try to experience the sacred from behind a desk. It's better, he says, to get out into the mystery of nature to find it. *Mejor ver y conocer el sagrado.* It's better to see the sacred and know it from experience.

As we head back to the house at Campo Flores, I think about the sacred delta Don Onesimo has shown me—the corvina in the river, the *biznaga* in the mountains. And with his explanations and stories, he has shown me something no one else could.

A good friend of mine, Douglas Burton-Christie, teaches theology at Loyola-Marymount University in Los Angeles. Doug has written extensively on how we discover the sacred in nature, how we rediscover the mystery of the tangible world. "Ordinary but necessary," Doug writes in one essay:

> Perhaps that is as good a place as any to begin rethinking what we mean by sacred. More and more in contemporary discourse the sacred seems to refer to precisely this: that which we cannot live without. The ground of being, to use an ancient philosophical term. . . . The poet Wallace Stevens refers to [it] as "transcendence downwards". . . a sense of the wholly mysterious Other who beckons to us through a world of infinite grace and beauty. This world. (Burton-Christie 1999, 48)

For Don Onesimo, the delta itself is the necessary, the sacred, thing. It is necessary because it feeds the Cucapá: gives them fish, gives them fruits of the desert like the flowers of *biznaga*. And it speaks to them in their sacred stories, grounding their lives with mystery and meaning.

Don Onesimo and Monica are fighting to protect the physical delta, the world of fish and food. But this material world, *lo material*, is inseparable for them from the sacred delta, *lo sagrado*. This sacred world resides for them in their own spiritual geography of the delta, which I was privileged to glimpse in the cactus garden.

The name of this place is *ba wi mok*, Don Onesimo says. It means, water on the other side of the mountains. We speak in Spanish but he also gives me the Cucapá name of all the places we visit. We are standing in the place where the Cucapá believed their dead went to live in the afterlife. It was considered a fertile place, green and wet and full of wildlife.

But you would not know that from what we see—neither water nor wildlife. We are standing in a blistered and blinding wasteland of salt. In places the salt is six inches deep and it covers the whole below–sea level depression. Fifty miles long, ten miles wide—a basin of salt. In Spanish it's called *Laguna Salada*, the Salty Lagoon.

Some early whites who visited this spot in the delta—and they were very few—called this the most inhospitable place in North America, as daunting as Death Valley and the nearby Salton Sink. Temperatures in the summer daily top 120 degrees Fahrenheit. On one side, the Sierra Cucapá hovers in blue haze. Farther to the southwest, forming the other boundary, is the Sierra Juarez. Laguna Salada lies between the mountain ranges, angling northwest in a long sunken branch off the main stem of the delta.

Strangely, we have come to a fishing camp in the middle of this salted plain. Don Onesimo walks out across the lonely salts. I join him. Its crusty surface cracks beneath our feet. Beneath inches of salt, we sink into a still-moist mud. It is rich and organic. There is the faint odor of dead fish. Don Onesimo points to long gouges cut into the surface of the salt. The keels of fishing boats made them, he tells me.

Shunned by whites, this salty spot is one of the Cucapá's favorite and most important fishing grounds. When the river runs high and the sea tides swell, the water pours into this low lagoon through a pass between the two mountain ranges. It filled with water as recently as 1997 and 1998, when heavy floods were flushed down into the delta. The fishing boat tracks we see now in the salt plain are from the time of that flood.

Every fish in the delta region can be caught in this shallow lake, mullets and corvina and everything. "*O, sí, un montón de pescado,*" says Don Onesimo. A big heap of fish. When the basin fills with water, it is completely transformed. It becomes not a desert of death and salt, but a haven for wildlife and the Cucapá. Osvel tells me that the U.S. Fish and Wildlife Service did a fly-over census of waterfowl in the basin during the last time it was flooded out. In a single pass, they counted over fifty thousand ducks.

Yet like everything else in the lives of the Cucapá, the Laguna Salada has become a bitter example, a symbol of a century of loss and persecution. For decades the Cucapás' lands were taken from them, including the Laguna Salada. In the 1930s, then-Mexican President Lázaro Cárdenas redistributed land as part of his agrarian reforms, instituting the *ejido* as the basic farming unit. The Cucapá were put into an *ejido*. It failed miserably. After years of fighting, in the early 1970s the Cucapá were finally given land.

Their territory includes the Laguna Salada. But Don Onesimo insists the land is largely worthless. None of it can be cultivated. The area is all down in the intertidal flats or tamarisk stands, salted-out lands worthless for farms. And the Cucapá were once a farming people. When the Laguna Salada does occasionally fill with water, it becomes a cruel joke to them. By law, the Mexican government controls all water in the country, so the Cucapá do not have the water rights within their own lands. The government sells fishing permits to Mexicans to fish in the Laguna Salada. This situation is a deeply bitter thing to the Cucapá—they lost their rights to fish in their accustomed grounds, and "Mexicans get to fish in our lands," as Don Onesimo puts it.

As we drive back to El Mayor, through the treeless Sierra Cucapá, I am thinking about the treatment of the Cucapá. They have no water rights. Their numbers are declining. They do not have water for even basic survival purposes. Yet just across the border, only an hour or two away, water is used for startling luxuries.

The Yuma Golf and Country Club, for example, uses 1,025 acre feet of water per year. That's more than 34 million gallons of water. An acre foot could provide water for a family of four Cucapá for a year. In Arizona there are 292 golf courses, all of them using water to keep the greens and fairways in peak condition. In Nevada, there are 88 golf courses. In Palm Springs, the Marriott Hotel has built an indoor lake that requires more than 50 million gallons to fill. Sun worshippers by pools can be cooled by sprays of soft mist. Flamingos even have a sprinkler for themselves. These are only examples of the kind of water waste in which Americans indulge.

In an eloquent review of water issues from around the world, *Pillar of Sand: Can the Irrigation Miracle Last?*, Sandra Postel argues that water is emerging as one of the central social and environmental problems all over the globe. The problem, she states succinctly, "has been clear: How can we meet the growing human needs for water without destroying the health of rivers, lakes, and other aquatic systems that we depend upon and that provide so many benefits?" (Postel 1999, viii).

After reviewing circumstances around the globe, Postel says the issues took on their true meaning for her when she visited El Mayor and the Cucapá people in Mexico. The Cucapá showed her that the problems with water are not technical problems any longer, as they were once considered, though technological solutions are needed to help solve growing problems of water scarcity and water quality. Water problems are not, she argues, even primarily legal, though more enlightened legislation is required.

Most fundamentally, she says, the plight of the Cucapá demonstrates the need to rethink our relationship to water. The Cucapá drive home for her in a visceral way the need for what she calls a new "water ethic." For her that ethic would involve two principles: self-sufficiency and sharing. She explains these ideas in terms of the Cucapá Indians of Mexico, and she states that such a water ethic would transcend the legalisms that allow North Americans to avoid taking responsibility for the condition of these Mexican Indians:

> Had a guiding ethic of self-sufficiency and sharing been adhered to when the Colorado River was allocated, enough would have been provided to sustain the Cocopa [sic] people and the ecosystems of the Colorado delta before water went into luxury activities upstream. (Postel 1999, 262)

Perhaps you could argue that when the original treaty was signed in 1944, no one really knew about, or recognized, environmental issues with regard to the river. Or you could say that the social issues were Mexico's problem, as American negotiators sought to maximize the United States' share of the water. Neither Don Onesimo nor the Yuma clapper rail counted in that different historical era. Still, such explanations and excuses would have been inappropriate even then. But they most certainly do not apply now. Surely we know too much, for example, to simply ignore the problem while we play another round of golf.

🌿

As we drive down out of the mountains, Don Onesimo and I, we have an experience that provides a vivid example of a different ethic in our relations with nature and with water. Toward the base of the rocky track through the mountains, we spot a coyote near the road. It sits on a pile of talus that juts from the base of the mountains, almost like tailings from a mining operation.

We stop the truck and get out.

The coyote trots away and is joined by its three packmates. At a safe distance, they stop to watch us warily. In the paloverdes and resting on the talus slopes, a flock of twenty or so turkey vultures skulk about the edges of the scene. Not far away, we discover why the coyotes and vultures are hanging around.

*In Cucapá stories, Coyote carried embers of fire in its tail and showed the people where they could find water in the desert mountains.*

They are feeding on a cow carcass.

Classic opportunists, the coyotes do not want to leave this cow, even when we stop to watch them. We wait and soon they begin to return to feed, cautiously, hanging back, but drawn to the meat. Cows do not do very well in this marginal land. Not many are here. There are as many skeletons of cows, I think, as there are living cows.

As we watch, Don Onesimo tells a story about Coyote, one of the Cucapá's culture heroes. He says that Coyote brought fire to the earth, carrying a coal from the sun in the tip of his tail. He says that Coyote even stole the heart of the creator, Sipa, and ran away with it someplace very near the pass in the Sierra Cucapá that we had just crossed. He tells me as well that Coyote had discovered freshwater in a spring near the Laguna Salada. Coyote also showed the spring to the people. That's why they call the spring *Pozo del Coyote,* Coyote Springs, he said. "It's not *Pozo de Agua,* as the Mexicans say," Don Onesimo insists energetically. *Pozo de Agua* means the "Water Well." The Cucapá name instead honors the spring's original discoverer, who shared what he found with the people.

The story of the Coyote teaching the Cucapá where to find water—the two species learning to live together—bespeaks a Cucapá ethic of generosity and inclusion. These people were always widely regarded for their willingness to help whites in the deserts, for example. Coyote shows them water and gives them a secret to life in the part of the delta that whites consider to be one of the most hostile on the continent. There is an environmental ethic contained in the narrative, an ethic of sharing water that embraces both people and animals. The beautiful Spanish word for this ethic is *convivencia,* which means "living together."

"So I don't kill Coyote," Don Onesimo declares, "even if one comes on the patio of my house."

Don Onesimo and I climb back into the truck and leave the desert to the coyotes and the vultures, and their dead cow.

# PART II

# Delta Water and the River of Law

*In human history, we have learned (I hope) that the conqueror role is eventually self-defeating. Why? Because it is implicit in such a role that the conqueror knows, ex cathedra, just what makes the community clock tick, and just what and who is valuable, and what and who is worthless, in community life. It always turns out that he knows neither, and this is why his conquests eventually defeat themselves. In the biotic community, a parallel situation exists. Abraham knew exactly what the land was for: it was to drip milk and honey into Abraham's mouth. At the present moment, the assurance with which we regard this assumption is inverse to the degree of our education.*

—Aldo Leopold, "The Land Ethic," in *A Sand County Almanac*

# "A Bridge over Troubled Waters"

In CADILLAC DESERT, author Marc Reisner writes that the Bureau of Reclamation is "the world's largest amalgamation of engineering talent" (Reisner 1993, 235). It is also a truism that the Colorado River is the most heavily litigated river in the world. A joke among the "water buffaloes" and "water hogs" of the river, what you might call the river establishment, characterizes the state of affairs with nice wit. There are only two species in the river's entire watershed that are not endangered: engineers and lawyers.

The Colorado River is so important to life in the arid West that the armies of engineers and lawyers who specialize in protecting rights and delivering water to the users have created a virtual separate river, a river of law—a thoroughly textual reality that has subsumed the literal reality of the river. The Colorado is codified in a body of law referred to as the "law of the river," out of which the arid West has made itself a river nation. If you don't "speak water"—terms like "cubic feet per second" for engineers and "present perfected rights" for lawyers— you will quickly feel like a stranger in a strange land at any gathering of these people. The textual nature of this second river cannot be overstated or ignored. It is the way the real river has been managed, controlled, and converted into a thoroughly human phenomenon throughout the post-dam years.

In a famous essay in *Harper's* magazine called "The West against Itself," the great historian, Bernard De Voto, described the West as a region in conflict with itself. One can use a similar paradox with the Colorado River. It is a river against itself. The river of law has transformed the river into a parody of its former self. As environmental concerns bring the real river increasingly into the national discussion about the management of the Colorado, the powers that be have surrounded themselves with the formal structure of law and text and have

◄ ◄ *Once a spectacular delta of wetlands and forest groves, the delta has been transformed in a century into one of the most fertile farmlands in the Western Hemisphere.*

◄ *One of the most endangered landscapes in North America, the Ciénega depends on decisions made by water managers in the United States.*

▲ *Monica González, of the Cucapá Natives of Mexico, is one of the leaders as communities in the delta organize to win water rights.*

resisted including the river's delta in the discussion. The law of the river has turned the river not only against itself. It has made the Colorado a river against its delta.

So complete is the law's control over and mediation of the actual river, that the only way to have an impact on the management of the river—its water—is to swim through this second river, the river of law.

Those who have benefited most from this body of law—and its benefits have by no means been equitably distributed within the United States or with Mexico—tend to capitalize the phrase: Law of the River. The style suggests the almost biblical status of this body of law for them. The metaphors used for discussing the law on the Colorado River also carry epic biblical echoes—people speak of "dividing the waters," for example, evoking Moses's forty-year journey through the desert in search of the Promised Land. Unconsciously such comparisons raise the law of the river almost to the status of the Ten Commandments, handed down by God as Americans imagined themselves the "chosen people" creating their own civilization in the desert.

The "water fundamentalists"—states, cities, water districts, irrigation districts, and others—tend to forget, however, that this second river of law is a thoroughly human text, not a divine one. As with all law, it is a product of history, changing over time, reflecting evolving social values and cultural imperatives. It was, in fact, born in contention. Its history has led its most respected historian, Norris Hundley Jr., to call the Colorado the "River of Controversy."

Now the river of law seems to be entering into a new round of controversy, a new round of "water wars." Out of a largely agricultural past, a New West has emerged, urban and heavily populated. The growing sentiment in favor of environmental protection generally and the Mexican delta specifically has led a swelling number of people to search for ways of bringing this river of law into greater alignment with emerging social and ecological values. These advocates are brought up sharply against the deeply vested interests and the contentious history along the Colorado, and as they fight for a new vision, the delta has emerged as increasingly central in the effort to bring an enhanced ecological emphasis to the law of the river. That effort has been inspired by the partial recovery in the Mexican delta itself. Since it happened in large part by accident, many would like to see legal mechanisms developed that would assure the current levels of restoration and, later perhaps, enable more. Also, the Mexican delta offers some of the best opportunities for genuine wetlands restoration on the river, and perhaps even on the continent.

Finding the legal mechanisms that would save the delta is the subject of the second half of *Red Delta*. The journey in this section is the search for a successful way to more effectively connect the delta with the river itself above the international boundary—leading to a transboundary, unified approach to this watershed. That means making the Mexican delta part of this second river, this river of law. Little is possible, especially with regard to the environment either in the United States or Mexico, without reference to this law. Yet the law of the river is almost entirely United States law.

Changing the relation of the delta to its river is a complex effort, largely because the law still has relatively little provision at this point either for the environment itself or for transfers of water to Mexico for ecological purposes. The challenge is no less than the search for a legal bridge over the Colorado River's troubled waters.

Lawyers and advocates for the river have focused primarily on two strategies to date. The first is largely domestic in its reference, and involves the application of the Endangered Species Act, charging that the act's jurisdiction for federal agencies like the Bureau of Reclamation is not limited to national matters, but applies across international boundaries as well. The second is largely international in its reference, and involves diplomatic and cooperative binational negotiations to develop an amendment to the 1944 Water Treaty that governs the allocation and deliveries of water to Mexico.

These binational environmental initiatives face several challenges. The first, and not the least of them, is the law of the river itself, which places human consumptive uses as its highest priority. For a long while, arguably to the present, human consumptive water use for agricultural, industrial, and municipal growth was the law's almost exclusive priority. Historically, the environment and wildlife have always come last in the law. Additionally, water managers and bureaucrats in the United States have denied any responsibility for environmental impacts or ecosystems in Mexico, claiming that these problems are largely a matter of Mexican sovereignty and Mexican remedy.

There is the overarching problem of limits in the actual river. The law of the river has already promised more water to users than the river actually carries. Everyone emphasizes this problem, and it fuels much of the bitterness and animosity of the new round of water wars emerging in the West. People are growing increasingly jealous of the limited amounts of water in the river. This problem is so severe that it threatens to swamp the entire effort to save the Mexican delta. To preserve their own vested interests, states and cities and big water users have massed their armies on opposite sides of the river and look ready to fight. In the middle of this battleground sits the Mexican delta.

It is important to note that so far the overallocation has taken place only in the second river, the river of law. Only three of the seven basin states actually *use* all (and in California's case, more than) their full legal allotments (Arizona has not yet actually used its full share, but the Central Arizona Project now makes full consumption possible). There is at least some surplus water in the system, though it is already being fought over. The question ought to be, should that extra water support yet more development in the Southwest, or should it be made available at long last for ecological restoration and social equity?

The Mexican delta offers an opportunity for redefining our priorities with regard to water. The following pages describe the amazing efforts of advocates of the delta in Mexico to change this second Colorado River—the law of the river. More deeply, these efforts point the way toward finally developing more sustainable patterns of water use in the West. These are struggles to develop, at long last, a healthier and saner relationship to the river, its water, and all the children it supports.

The fundamental premise of the pages that follow is that if we care enough about the ecological health of the river and its delta, we can find the legal mechanisms that will ensure its continued conservation. Complicating issues is one of the most effective strategies used by the water fundamentalists: bogging discussions down in minutiae. The high priests of this water nation have used "water speak" to exclude people, making the discussions impossibly technical, and making participation available only to the initiates of the faith.

It is easy to get lost in the details. But it is crucial to keep one's eyes on what is at the heart of the whole debate—bringing the river into greater harmony with itself and its delta.

*The fertile sediments of the delta have been transformed through irrigation from the Colorado River into one of the richest farming regions in the Western Hemisphere.*

169

# Delta of Desire, River of Law

Northern harrier and least bittern
PLACE: Confluence of the Río Hardy and Colorado River

THE WATER IS RISING. At our feet at the boat launch, we can see the waters creeping higher. Rumor has it this morning that the Colorado River is flowing again to the sea. Lake Havasu behind Parker Dam in the United States is being drawn down.

The water at our feet is a powerful example of life in the delta. Lake Havasu is being lowered for "dock repairs," according to the report on the radio. But it also means water is flowing again in the Mexican delta. Water decisions in the United States, even for such minor matters as docks, shape water realities in the delta.

For Octavio Schlemmer and German Muñoz, this change in flow gives a great opportunity to explore the river. They suggest we go out on Río Hardy, head south for the sea, and look for the confluence of the Hardy and Colorado Rivers. We jump at the chance. The two lawyers I'm with clamber into the boat and take their places in the middle. I sit in the bow while Octavio and German launch us into the Río Hardy.

"Let's go find the Colorado River," says Bill Snape, Defenders of Wildlife's head lawyer. Claudio Torres Nachón smiles, but says nothing. Claudio is a top young environmental lawyer in Mexico. He is with the Centro de Derecho Ambiental e Integración Economica del Sur, A.C., in Veracruz, Mexico (Center for Environmental Law and Economic Integration of the South).

We crowd into a small aluminum boat on the pesticide-laced Hardy River, surrounded by weedy tamarisk. It is a toxic scene on corrosive green waters,

*◄ A great blue heron takes off against the luminous light on a delta pond.*

*▲ Rarely seen, least bitterns fish from deep in the reeds of the Ciénega.*

171

enough to make anyone feel murky and mad. Yet I realize that Octavio and German view the visit by these two lawyers as an occasion for hope. For good reason, too. In the litigious world of Colorado River water politics, nothing gets done without a lawyer.

These two men are doing everything they can to treat our small entourage like visiting royalty on a tour of the river.

Here in the lower delta, Octavio and German know that these two visiting lawyers are a special event. In their wildest dreams, they might never have imagined that two of the most powerful environmental lawyers in the United States and Mexico would come to see them. For decades now Octavio and German have been among the victims of water policy on the Colorado River. Octavio is a small, quiet man of German and Mexican descent. He has lived along Río Hardy since 1949, over half a century. In that time, he has watched Río Hardy and the delta deteriorate around him. German (pronounced "Herman") is tall and has an oval face. According to friends, German is the last man you might have thought would become an environmentalist. But after watching his tourist camp business on Río Hardy crash in the last twenty years, he has realized that his economic health depends on the river's health.

We launch the boat from German's camp, Campo Muñoz. It's much like Don Onesimo's enclave, Campo Flores: dreary buildings in a clearing amid acres of tamarisk by the river. Once Americans used to hunt at his camp, German says. Not many come any more, though there are a couple of American dove hunters at the ramp as we shove off.

Octavio and German are among the more than 200,000 people who live in the 1,127 small communities in the delta. About 5,000 people actually live inside the biosphere reserve, in communities like Indiviso. They are all water users. Like Octavio and German and the Cucapá of El Mayor, most have been left out of current allocations of water, which favor big agriculture and cities on both sides of the border. These men know that if they are to dare to dream a new delta with a restored Río Hardy, it will only happen by changing the laws that govern the flows of the Colorado River. And by getting help from people in the United States.

The delta communities have already begun to help themselves. Octavio and German are members of a new organization of users in the Río Hardy part of the delta, called Asociación Ecológica de Usuarios del Río Hardy y Colorado (AEURHYC). In community workshops sponsored by organizations like Pronatura, in Sonora, the group has begun to speak out in Mexico and to develop an action agenda. Claudio Torres Nachón is here to talk with them about legal action in Mexico. In addition to the successful human rights suit on behalf of the Cucapá, Claudio has brought a *denuncia popular*, or complaint, on behalf of these and other people, claiming the government has neglected care of the natural environment in the delta. Octavio and German are eager to meet with Bill Snape because their group is part of Defenders' major lawsuit inside the United States under the Endangered Species Act. Lawyers and activists on both sides of the border are building cooperative networks.

"In this area, Río Hardy is the source of life," says German. *La fuente de la vida.* "We know we have something very valuable. We know it's damaged. But we don't know what we should do."

Apart from the sickening cast of the water of Río Hardy, the river is small, about fifteen yards across. German gestures. Once the river was wide, he says. It was deep too. But now, he tells me, sometimes the Hardy has almost no water. "Two fingers deep sometimes," he says. *"Una lamina de agua, un espejo de agua,"* he says. A thin sheet of water, a mirror of water.

The tamarisk on both sides of the river is so thick and so dominant, Octavio explains, that it can't be eradicated. They have tried pulling it up. It always comes back. More seeds just float down on the rivers. He hates the tamarisk.

*"Una plaga,"* German sneers. A plague.

We pass several more derelict and dilapidated *campos*. Almost all look abandoned, like German's Campo Muñoz, with the occasional Mexican living in one of the houses. When the sandbar blocked Río Hardy, this used to be a paradise in the delta, German explains. Fed by springs as well as the geothermal runoff from the north, a huge marsh of startling dimensions formed—fifty thousand acres of some of the richest wetlands in the Southwest, the Río Hardy Wetlands. They must have been a remarkable refuge for endangered species like the Yuma clapper rail in those years, before the Ciénega de Santa Clara existed.

"We had lots of ducks, shorebirds, doves. The hunters loved to come here from America. We would take them into the marshes," German says. The thousands of acres of tamarisk have destroyed not only cottonwood and willow forests in the delta, but the invasive salt cedar has displaced the mesquite and quailbush and arrow-weed that once made a dry forest around the riparian zones.

The same floods from the dams upriver that forced the Cucapá to move out of their river houses and into El Mayor destroyed these wetlands. When the wetlands vanished, when Río Hardy shrank to a phantom river, and when the tamarisk replaced the native trees, the Americans disappeared, and with them, the economy.

"Now there are no ducks to hunt," German says. "No fish. No tourists. The Americans won't return to their ruined houses."

Octavio and German are now allies with the Cucapá in the effort to restore these wetlands. Together with others in AEURHYC, they hope that the sandbar can be rebuilt, perhaps an earthen dam, and the wetlands returned.

*Lawyers Bill Snape of Defenders of Wildlife (center) and Claudio Torres Nachón of Mexico (left) receive a tour of the polluted Río Hardy. Octavio Schlemmer (right) represents the association of water users along the Río Hardy.*

Claudio Torres Nachón, the Mexican lawyer, is shocked by what he sees along the Río Hardy. He is visiting this part of Mexico for the first time. He and Bill are meeting to discuss their interest in coordinating legal action for the environment in both Mexico and the United States.

"It's a failure for you not to have more water to continue your way of life," Claudio tells Octavio and German in Spanish. "These are both environmental and human rights violations."

After this visit, Claudio's environmental organization filed suit in Mexico on behalf of the Cucapá and AEURHYC. This action is the beginning of legal pressure on the Mexican government to address the environmental issues in the Colorado River delta. The suit is also coordinated with legal efforts of Defenders of Wildlife and other environmental groups in the United States and Mexico, specifically the suit under the Endangered Species Act.

The ultimate goal of these legal efforts is to get more resources for both social and environmental justice. The fact that environmental lawyers from both countries were touring Río Hardy signifies the binational dimension of the cause—both countries are implicated, both countries must begin to place the delta's environment as a priority, both countries are responsible for what has happened to the people, the plants, and the creatures of the delta in Mexico.

"We believe there's nothing in the ESA," Bill tells the two Mexicans in the boat, "that suggests its reach on behalf of the environment should necessarily stop at the U.S. border."

This is potentially one of the most powerful and precedent-setting suits under the Endangered Species Act to be filed since the act was passed in 1973. "The ESA is one of the few pieces of legislation in the world that has real and direct clout for specific habitats," Bill says.

Even as we are boating down Río Hardy, we are aware that to live in the Mexican portion of the delta is to live in some sense at the mercy of actions taken in the United States. The river is rising—and these Mexicans have had nothing to do with it. Americans made that decision. To U.S. managers, it is a simple hydrological decision to draw down a reservoir. To us in the delta on the rising Río Hardy, it is testimony to the interconnection between the two halves of the lower Colorado River, U.S. and Mexican.

We cruise slowly down Río Hardy, talking about its history and its destruction. Then we spot the Colorado River. At first we cannot see the river. We know we are approaching it because the trees change. The tamarisk gives way to a thick clump of reeds, tall bulrushes called *carizzo*. The plants are a sudden splotch of bright green in place of the blue-gray of the tamarisk—a sign of freshwater feeding into the pesticides of the Hardy. A feeder stream enters into a bend of the Hardy, like a small tributary. The streams join and both head south toward the Gulf.

We turn up the feeder stream.

*"El Río Colorado,"* German says.

It's an unrecognizable river. About five yards wide, it is a brighter green than the Hardy. And much smaller. There is something pathetic about it, an old river hobbling to the sea, a powerful image of a river enduring its own living death.

"Behold the mighty Colorado River," Bill mutters.

Yet it is impossible not to see something other than fallen greatness in this impoverished Colorado River. It is flowing only by the grace, as it were, of the Bureau of Reclamation, but the mere presence of water in the river makes it clear that the river is not entirely dead, at least. This river water still supports wetlands and wildlife in the area, however reduced, hidden back in this channel. Carlos Valdés and Osvel Hinojosa found hidden wetlands near here, as well as endangered Yuma clapper rails and other species.

The waters of the Colorado also feed the dreams of Octavio and German, among others, for a restoration of these wetlands, for the return of more ducks. It is on behalf of this fallen river, and the creatures and people in Mexico that depend upon it, that Defenders of Wildlife and its coalition filed suit under the Endangered Species Act in the United States.

The lesson of the recent revival of the Mexican delta has been clear: if water is provided, the ecosystems will come back. The question raised is, how can that water be provided? The lack of water to the delta through much of this century has created the impression that the dams have choked up the river. That there's virtually no water available for ecological purposes in the United States, much less in Mexico. On top of this, the users in the United States guard their allotments from the river with more than jealousy. The joke among water lawyers in the West is that their clients' water rights are more valuable than even their own children.

This is where it is crucial to keep the distinction between the river of law and the real river distinct. No Moses-like miracle is required to divide the waters again in the Colorado River, though some deft legal tricks might be in order. What is required is changing the law to make it possible to re-allot the little water that remains unused—not unclaimed—in the Colorado watershed. The miracle that lawyers need to perform is not making water spring from stone, but rather making it spring from words.

"The Colorado River once carried a heavy burden of silt on its journey to the Gulf of California," writes Peter W. Culp in an excellent article on the legal framework governing the Colorado River. "It now carries an even heavier burden of law." The article is entitled "Restoring the Colorado Delta with the Limits of the Law of the River: A Case for Voluntary Water Transfers." It is an excellent place to find a succinct treatment of the new water wars in the West and of the efforts of environmental lawyers to secure water for the delta.

Perhaps these new water wars take the shape of a classic Shakespearean battle between desire and the law: desire for a new and more natural river, and the law that has hedged the river in a straightjacket of claims and concrete dams. In another useful analysis of the law of the river, David Getches of the University of Colorado Law School argues that the law needs to be changed for the sake of greater equity, efficiency, and environmental sustainability. The states, he says, may resist changes to the law as it now stands. And they will do so for good reason. Most of those with vested interests have been the main beneficiaries of

*Against all odds, the noble Colorado River still flows in the Mexican delta. But it is a diminished and degraded version of itself as it stumbles into its confluence here with Río Hardy.*

current law—and they have received "massive federal charity" in building water projects that they could never have afforded. Yet Getches argues that the history of the law of the river is that it has created clear winners and losers, haves and have-nots. Not everyone or everything has shared equally in the wealth created out of the damming of the river. Many have been excluded even from participation in the allocation decisions. Most notably, he says, Mexico has been excluded, as have Native American tribes, and, needless to say, the environment.

"The most pervasive problem in the governance of the Colorado River, then," writes Getches, "has been the exclusion of diverse values and views." He goes on:

> Having grown out of a preoccupation with allocating rights to consume Colorado River water, the law of the river ignores the wider range of values that people in modern society hold for the Colorado River. The consequences of their exclusion are manifest in economically wasteful, politically inequitable, and ecologically unsustainable uses of natural resources in the Colorado River basin. Some commentators blame the present condition of the Colorado River on existing laws and urge fundamental changes. (Getches 1997, 576–77)

Increasing water flows to the Mexican delta of the Colorado River is becoming the single issue around which these new emerging values for both social equity and environmental sustainability have coalesced.

Defenders' lawsuit also highlights the ways in which the law of the river must adapt to new values and a changing appreciation of the actual river. To understand how this lawsuit works, you must first understand a bit about the law of the river—and whom it benefits.

We will try to avoid the technical obfuscation and impenetrable legal jargon that clouds the river of law.

The law of the river controls the way that water from the Colorado River is distributed. How? Agriculture has been and remains the primary beneficiary of the water in the river. The numbers are changing as population in the arid Southwest shifts. As of 2000, the Bureau of Reclamation estimates that 77 percent of the water in the Colorado River goes toward farming. Urban users are increasing in their consumption of water from the river in recent years, with about 30 million people in the U.S. portion of the Colorado watershed dependent on the river. Urban uses—cities and industries—account for 23 percent of the river's allocation. The rapid growth of urban use of the water accounts for increasing human pressure on the Colorado.

Up until now, the major beneficiary of the damming of the Colorado River has been not human beings, but cows. "Cows are far and away the chief beneficiaries of water, even in the West's most populous regions," writes Philip Fradkin Jr. in *A River No More*. Of the 99 million acres along the lower Colorado River, for example, 82 million are used as rangeland or pasture for livestock, he writes, turning the river's watershed into "a vast feedlot for livestock." The dominant crop is alfalfa for cattle feed.

"Probably never in history has so much money been spent, so many water-works constructed, so many political battles fought, and so many lawsuits filed to succor a rather sluggish four-footed beast," he concludes (Fradkin [1968] 1996, 31–2).

Yes, you might say, but cows provide food for people—hamburgers. But hamburgers are a water-expensive way to get food. Consider this: It takes about 2,600 gallons of water to produce one pound of beef. It takes about 12 gallons of water to produce one head of lettuce. You don't have to be a vegetarian to realize that just cutting back a bit on meat production could save a lot of water.

The law of the river is the legal framework within which the water is dispersed and allotted. It is not a single document. It is not even a coherent body of law. It grew up over the last century and is a complex and even confused mire of interstate compacts, international treaties, administrative regulations, court decisions, federal laws, state laws, contracts, and unwritten understandings.

"It's more a set of religious than legal documents," Robert J. Glennon, Morris K. Udall Professor of Law at the University of Arizona, says. "Despite its mytho-logical status, it's really a motley collection of contradictory, ad hoc, and vague legal positions."

Despite the laws' almost numbing complexity, there are essentially three major sets of documents at the heart of the law of the river:

The 1922 Colorado River Compact and subsequent 1928 Boulder Canyon
    Project Act
The 1944 United States–Mexico Treaty, with its amendments and "minutes"
The 1963–64 *Arizona v. California* Supreme Court Decision and Decree

There are many other documents of greater or lesser importance, depending largely on where you live and what you do for a living. In more recent times, a whole body of environmental law has also emerged and is clearly a part of the law of the river. But its exact relation to all this earlier law is still being worked out. Plus, it should be said, few lawyers actually agree among themselves on exactly which documents are the core of the law.

According to Peter Culp, author of the article on the delta and the law, at the heart of the law of the river is a single, unifying concept: "beneficial consumptive use." The exact legal meaning of this phrase is not even clear. It is nowhere defined, Culp notes, and anyway, he claims it is an "archaic notion"

(Culp, 2000, 14). But it is a notion that is implied throughout the central document of the law of the river, and it does make clear that the water in the river is for people to use for their purposes.

Notions of what it means to "use" the river vary in different legal documents in this body of law. The 1922 Colorado River Compact gives an implied definition of beneficial use as "water applied to domestic and agricultural uses." Domestic use "shall include the use of water for household, stock, municipal, mining, milling, industrial, and other like purposes" (quoted in Culp 2000, 14).

The Colorado River Compact of 1922 and the subsequent Boulder Canyon Project Act of 1928 were the first legal documents to address water allocation of the Colorado River. Many consider the agreements separately, but they are more useful when considered together. The original compact grew out of anxiety over California's massive use of the Colorado in the first years of the last century, beginning in 1901 with the diversion of the river to the Salton Sink, renamed the Imperial Valley, and the subsequent selling off agricultural and water rights. The scale of these diversions, and fear of California water greed, awakened the other states along the river and prompted them to enter into negotiations that would protect their claims to the water in the river. The highly respected Secretary of Commerce Herbert Hoover was central in convincing the warring states to compromise.

The compact divided the Colorado River in the U.S. portion of the watershed into two basins, divided at Lee's Ferry, Arizona. The Upper Basin includes four states: Colorado, Utah, Wyoming, and New Mexico. The Lower Basin includes the three states along the desert portion of lower river: California, Arizona, and Nevada. The compact also agreed on a rough disposition of the water in the river between the two basins: 7.5 million acre feet (maf) to each. Another 1 maf was assumed to be available to the Lower Basin from tributaries in that region. In total, 16 maf were allocated from the river.

The Boulder Canyon Project Act is a piece of legislation passed by Congress that confirmed the earlier compact and added several key provisions. It divided the allocation among the three Lower Basin states and it authorized the construction of the Boulder Canyon Dam, later renamed Hoover Dam in honor of the role played by Herbert Hoover in apportioning the river. The allocations are as follows: California, 4.4 maf; Arizona, 2.8 maf; Nevada, 0.3 maf.

The Boulder Canyon Act also authorized the construction of the All-American Canal, which delivers water to California's Imperial Valley along the international border, avoiding the need to move the water through Mexican territory.

Even after the act was passed, the Lower Basin states continued to dispute the proper allocation of the water in the river, which led to decades of litigation. As a result, the secretary of the interior was named "watermaster" of the Lower

Basin, with the power to make a number of key decisions on the way water gets divided—particularly concerning surplus waters in times of heavy flows.

Lagging behind in development, the Upper Basin states did not settle the division of their portion of the waters of the Colorado until 1948. The Upper Basin Compact of 1948 gives each state a percentage of the river's water determined by the contribution each state makes to the river's flow.

The second and truly major document in the law of the river is the 1944 United States–Mexico Water Treaty (henceforth called the 1944 water treaty). This treaty is also at the heart of the effort to find more water in the Colorado River for the Mexican delta. The key provisions of this treaty, as regards the Colorado River (the treaty also pertains to the Río Grande along the Texas–Mexico border), are as follows: Mexico is guaranteed 1.5 maf of river water. In surplus years—when the river runs high with excess water from heavy snows and rains—Mexico can receive an additional 200,000 af. Additionally, Mexico is guaranteed first priority in receiving its allotment of water from the river. Before any other water is distributed in the United States, regardless of drought or other problems natural or manmade, Mexico gets its water.

In rough percentages, Mexico is guaranteed about 10 percent of the flows of the river. When the treaty was signed, on February 3, 1944, it may have looked like a decent deal to the Mexicans. After all, the treaty had taken some eighty years of intense and bitter negotiations. But from the perspective of history, it is clear that Mexico was improvident in signing the treaty for so little. Why it agreed to so little of the river is one of the great, unanswered historical questions. Some have doubted the skill of the Mexican negotiators. Others have blamed the bullying tactics of the United States. Some have even gone so far as to claim that the United States essentially "stole" the river from Mexico.

As the leading historian of the 1944 water treaty Norris Hundley Jr. makes clear in *Dividing the Waters: A Century of Controversy between the United States and Mexico*, the United States did not reach out to include Mexico in plans to develop the Colorado River. In fact, Mexico was excluded from discussions about developing the river, just as they are now excluded from participating in discussions on how the river is managed. Negotiations in the treaty proceeded by fits and starts over the decades, in an atmosphere of profound mutual distrust, and many of the talks took place during some of the most turbulent years—including revolution and civil war—in Mexico's history. The states were intensely opposed to sharing the river with Mexico. They may have been fighting among themselves over the allocations. But they could all come together to agree on one thing: Mexico was not entitled to even one drop of water. The doctrine of Attorney General Judson Harmon was readily taken up by the states. He claimed that the United States had no responsibility of any sort, under international law, to share any of the Colorado River. The Harmon Doctrine was repudiated by jurists even in the

United States, but it defined a prevalent and even dominant attitude on the part of those greedy for the wealth made possible by the river in the West. As Hundley writes:

> Placed thus by geography in a decided advantage over Mexico, the United States developed a selfish attitude that was reflected in the opinion of Attorney General Judson Harmon. (Hundley 1966, 40)

Mexico also pressed its case, claiming that it had rights to some 4.5 maf. The impasse remained for decades, a grand and tedious ebb and flow of frustrated diplomacy. But the United States needed a treaty with its "neighbor" during World War II, and besides, there were serious questions about the legal status of Hoover Dam when the "Mexican problem" remained unsettled. The federal government essentially forced the states to back down. A Mexican saying has it that many of Mexico's problems can be understood geographically: "So far from God, so close to the United States." Mexico clearly felt it was dealing with a much more powerful neighbor in the United States. What is clear is that Mexico settled for far too little water. Its 10 percent does not provide enough water even for its own agricultural needs, much less enough to cope with environmental issues.

The third major component to the law of the river is the 1963–64 Supreme Court decision in *Arizona v. California*. This major decision settled more than a decade of litigation between Arizona and California, the two most contentious and aggressive states in exploiting the Colorado watershed. Arizona wanted to build a major water project—the Central Arizona Project, or CAP—so it could exploit its full share of the 2.8 maf of the river's water. For arcane reasons that had more to do with California's insatiable greed for water than any reality—legal or otherwise—California tried through litigation to prevent Arizona from claiming its allotment. The Supreme Court rejected California's contention, opening the way for Arizona's CAP, which has recently been completed and which has formed the basis for a whole new round of water battles between the two states in the 1990s and continuing today.

The other crucial dimension of this Supreme Court decision was that it granted five Native American tribes along the Colorado River "reserved rights" to water, dating back to the establishment of their reservations. The Arizona Cocopah are one of these tribes. The Native American entitlements are to be met from the state in which the tribes are located.

Such is the heart of the law of the river. It controls the allocations among the major users. Among its great benefits to water users is that it gives both predictability and control to their relation with the water.

In addition to these laws governing allocation, we must look at the environmental law. If it is a serious historical question why Mexico got such a small portion of the river's water, it is an even more damning question why the environment was not factored into the law of the river at all until the 1960s. All the major dams on the mainstream of the Colorado River were built before there was a requirement for an environmental impact statement. When asked why the development of the river proceeded at such a breakneck pace without regard to the environmental consequences, most people respond that the major developments on the river took place before an environmental consciousness had emerged in the country.

But this answer will not do. By 1935, when Hoover Dam came on line, there had already been considerable environmental legislation passed in this country, including international treaties protecting migratory birds. As David Getches writes about the environment in his article, "Colorado River Governance," "The [environmental] losses were foreseeable. Naturalists and environmental activists warned [the river managers]" (Getches 1997, 596). Clearly, other factors along the river left broad participation by a range of people excluded, including Mexicans and U.S. environmentalists. This exclusion of environmental considerations in the period of dam building is another major historical question still calling for investigation and research—and answers.

Nevertheless, over the last several decades the law of the river has moved into a new historical stage. A number of environmental considerations have entered the picture along the fringes, making matters less clear and predictable for the future. They are not formally part of the law of the river—they have not affected allocations directly. But environmental concerns have pushed river management into a completely new phase, one that must somehow adapt to incorporate the new ecological consciousness that has emerged as the Era of Dams has ended.

The 1992 Grand Canyon Project Act, for example, requires the secretary of the interior to operate Glen Canyon Dam in such a way as to protect and improve the natural values for which the Grand Canyon National Park and Glen Canyon National Recreation Area were created.

The powerful pieces of environmental legislation that affect the law of the river are federal. The Clean Water Act and the National Environmental Policy Act (NEPA) offer environmental protection. NEPA requires federal agencies to consider environmental impacts of their actions. The most powerful new player among the new environmental laws along the river is the 1973 Endangered Species Act. It is the most comprehensive legislation for the preservation of endangered species ever enacted, with the potential to drastically change the way business as usual has been conducted on the river.

Despite competing interests, finding room for endangered creatures as well as cows along the Colorado River might be simply a matter of adjusting the law to reflect the emerging environmental consciousness. Endangered species like southwestern willow flycatchers and Yuma clapper rails could share water with livestock. Except for one problem.

The river is already overallocated.

The total amounts of water allocated by the law of the river for beneficial consumptive uses total over 20 maf, including evaporation: 16 maf among the basin states, 1.5 maf for Mexico, and 2.5 maf of evaporation per year. Tree-ring analysis by scientists at the University of Arizona have determined that over the last five hundred years, the annual flow of the river has averaged 13.2 maf. Even before the environment is factored in, the demands on the river are bloated by more than a third.

Into this closed world of law and scarcity, endangered creatures and riverine habitats are a powerful new concern.

Many people like to stress the overallocation of the river: It underscores a sense of looming catastrophe along the whole course of the Colorado River. The resources are stretched too thin, there's not enough water, how can we ever find enough to go around? So goes the lament.

It is a real worry—the demands on the river are intense. What is crucial to remember, though, is that it is not the real river that has been overused. It is the second river, the river of law, which is scarce. In fact, so far the Upper Basin states have not used their full share of the river, and excess water remains in the system. What is really happening is that competing interests are fighting now over this "excess" or surplus water—surplus, that is, as long as we have no drought.

So the vested interests resist the new players along the river. "It's a culture shock for many that they have to deal with us," Defenders of Wildlife's Bill Snape says. By "us" he refers to the environmental organizations who have brought the suit under the Endangered Species Act. "It's clear that the Endangered Species Act is a part of the law of the river. What we want to find out is, to what extent?"

And as one person in Arizona said to me when I asked how you get water from this river, there's only one way. "You sue," she said.

It's not that Bill Snape or anyone else wanted to sue. They only filed suit after a crisis of bad faith with river managers over the issue of endangered species management. At the heart of the controversy was the Mexican delta.

To find out why, I have come to the heart of the delta again. I am sitting on the concrete porch of a small, unfinished house in Ejido Luis Encinas Johnson, watching the dust settle around me. This is the *ejido* by the Ciénega de Santa Clara where Juan Butrón lives. I have been staying in this house; it has no running water, no amenities. But it is comfortable enough in a rugged sort of way. I sleep on a cot made of burlap strung between two-by-fours. The house is very dusty.

I have been waiting for another American to arrive this evening. He has agreed to come to the *ejido* from his home near San Diego, to meet with me and talk about the politics of endangered species on the lower Colorado River. His name is David Hogan. He is the River Program Coordinator for the Center for Biological Diversity in Tucson.

The main road through the *ejido* is a big dusty boulevard, with some struggling palm trees planted in the middle. I am watching trucks drive by as they kick up huge clouds of dust. Lesser nighthawks chase insects gathered under the lamplights. Life here is as slow as the dust settling in a fine film over everything.

Suddenly I realize that one of the trucks driving by is David's. He pulls up in front of the house and walks up to me with a huge smile. David has an imposing, impressive air about him. He is six feet six inches tall, has gray eyes, and a rectangular face. He describes himself as a "policy geek," though he does not look like any image of a "geek" I have ever had. Out of high school, he skipped college and went straight to the redwoods of northern California. With Earth First! he camped in the trees to keep them from being cut down. By the late 1980s, he returned to southern California to work on environmental issues related to the border. He discovered the Ciénega de Santa Clara near Ejido Luis Encinas Johnson, and it inspired him to work for the restoration of the Mexican delta.

Though he is involved in all the legal and political aspects of restoring the delta, I am particularly interested in talking to David about a program begun in 1995 on the lower Colorado River to address management issues for endangered species. The program is called the Multi-Species Conservation Program (MSCP). With a $4.5 million study over five years, the MSCP was to develop a plan to guide river operations over a fifty-year period, accommodating economic needs and protecting 102 species of rare plants and animals.

David was one of the representatives from environmental organizations who tried to take part in this program. It was a good-faith effort to help develop a region-wide approach to endangered species management that honored the Endangered Species Act in a cooperative, rather than confrontational way.

"It was a tale of intrigue that gave us no choice," he says to me on the porch. "We had to quit." What he means is that he and several other environmentalists entered the process in an open and cooperative spirit. They believed, perhaps naively, that their interest in the delta would be accommodated. They wound up feeling used and betrayed.

A large group of thirty-six representatives from state and federal agencies, irrigation districts, and water boards, established a steering committee of the MSCP. The Bureau of Reclamation is actively involved, as is the U.S. Fish and Wildlife Service. The website for the group states that ten seats have been

reserved for environmentalists. Four groups participated initially: the Center for Biological Diversity (represented by David), Defenders of Wildlife, American Rivers, and The Nature Conservancy. Other groups stayed out.

The stated goals of the MSCP make a lot of sense. It formed to develop a region-wide approach to conservation in the lower Colorado River. The website for the MSCP states that the program is trying to provide "habitat-wide zcompliance with the ESA." This "habitat" is defined as extending from Glen Canyon Dam to the Southern International Boundary—the southernmost point of the limitrophe between Arizona and Mexico along the Colorado River. Participants intend to develop a comprehensive recovery plan for endangered, threatened, and sensitive species in the region, hoping that economies of scale might coincide with more effective management strategies. The ESA requires that recovery plans be prepared for all listed species. The process can be expensive and controversial. Many conservationists, in fact, think it makes sense to focus recovery efforts not on individual species, but to take a broader, ecosystem approach.

In the case of the lower Colorado River, the major endangered species are four native fish: razorback sucker, Colorado squawfish, humpback chub, and bonytail chub. The endangered birds include the southwestern willow flycatcher, Yuma clapper rail, bald eagle, and peregrine falcon. Another two dozen species are listed as being of special concern, including northern harriers, a hawk of marshes and plains; the least bittern, a shy little heron of the marshes; brown pelicans and white pelicans; and the California black rail.

*David Hogan fought for the delta in the Multi-Species Conservation Program. The failure of gaining recognition for the delta has led to a major and ground-breaking lawsuit under the Endangered Species Act.*

185

"We got into the process, but we didn't have much hope," David says. "The question is, does river regulation include ESA compliance?" It began to appear to David and others that the MSCP process was a scheme to circumvent ESA compliance. Under the ESA, recovery plans must be developed for all listed species. Equally significant for ESA compliance, federal agencies like the Bureau of Reclamation are required under the law's Section 7 to actively promote and improve the well being of endangered species. If it is determined that agency actions will harm species, that agency can apply to the Fish and Wildlife Service for an exemption, but it must then "mitigate" for the affected species—that is, compensate for the harm by habitat restoration or similar positive actions.

In 1996, for example, the Bureau of Reclamation studied the effects of its operation and maintenance activities along the river. It concluded that four species would be affected: three fish (totoaba, razorback sucker, and bonytail) and one bird (southwestern willow flycatcher). The flycatcher, for one, was already determined to be in such bad shape that future operations could put a large portion of its population over the edge. And the totoaba occurs only in Mexico's Gulf of California: Yet the bureau refuses to mitigate on behalf of the totoaba. This lack of concern for the totoaba, which the bureau admitted would be affected, was ultimately one of the major reasons for the lawsuit under the ESA.

One of the foremost experts on Colorado ecosystems and freshwater fish, W. L. Minckley, wrote that the bureau's document continued a "three-decade-long pattern that we have personally experienced of attributing changes which result in endangerment of biota" to "things which have happened in the past," or "things that are the responsibility of other agencies," or "factors professed to be a result of the Bureau's mandated (legal) responsibilities." (quoted in Gillon, "Letter of Intent"). In other words, the bureau has consistently refused to take responsibility for harming endangered species.

David says he was especially worried about the flycatcher. "It looked like the bureau was just allowing the extinction of the southwestern willow flycatcher." This heralded the beginning of a "conservation crisis" for the MSCP. Environmentalists were poised to challenge the process with a lawsuit over this alone. "It looked like a way for the states and the bureau to get permits to go on doing what they've always done," David says. The MSCP process has been marked by a huge number of these intricate wrangles.

The crisis came to a head in 1998, though, over the Mexican portion of the delta. David says he and Defenders of Wildlife worked hard on this issue. "We raised hell," he says. "We told them they had to include the delta."

Since many of the listed species occur on both sides of the border, a "habitat-wide" approach to preservation would logically seem to include both U.S. and Mexican habitat. But the MSCP refused to consider its mandate as extending below the international border. The water managers have consistently denied that impacts in Mexico are relevant to federal decision-making (Culp 2000, 11).

"We made six or seven proposals," David says. "We tried to hash out a compromise with a couple of the sympathetic representatives. What we came up with was this. While the bureau studied the needs of a species in the United States, it would do a parallel study of the species in the delta [in Mexico]." The bureau would work unofficially with Mexico to get input on conservation plans.

"All we wanted was a good-faith look at the studies," David says. "The idea that the bureau and MSCP might actually allocate water for Mexico was not off the table, but not required either."

The matter came to a vote in the November 1998 meeting of the steering committee. "No supporters. None of the states would go with us," David says. "We'd been played. We resigned on the spot." All four environmental organizations dropped out of the MSCP process.

David is completely disillusioned. "The water and power interests along the Colorado River have a long history of endangered species and habitat destruction," he says. "They have no track record of wildlife conservation, and no interest in changing. The greed of the Lower Basin states is shocking." Meanwhile, the MSCP process is pushing ahead without environmental representation.

The environmentalists feel they were left with no choice but to file suit under the ESA. "We have this double strategy," says David. "We want to hit 'em over the head with a big stick, plus we have a carrot for what we want. The lawsuit is the big stick. A treaty amendment so that water could be transferred to Mexico for the environment is what we really want."

We leave off at this point in the conversation. All the kids have gone indoors. The already slow *ejido* has gone to sleep. That night David sleeps in the back of his pick-up truck. I sleep in the house, which Juan and others are trying to develop for their ecotourism business. The temperature is still near 100 degrees Fahrenheit, even at 10 P.M., a late-summer desert night. I open all the windows and turn the fan directly on me. The fan is partly to keep me cool, partly to keep the mosquitoes off me while I sleep.

I lie on a cot in the room, listening to the fan as it blows air on my skin. I think about how much more basic life is here in the *ejido* on the Mexican side of the border. One delta, I think, but two very different countries. Through the window, I watch the nighthawks and drift off, listening to them screech just outside the open window, as they whirl through the hot night, chasing bugs.

On November 18, 1999, Bill Snape from Defenders of Wildlife stood before a conference at the Araiza Inn, in Mexicali, Baja California. That he had been asked to speak on the MSCP process was one of the unspoken ironies of the conference. In the audience were all the major Mexican biologists working in the delta, as well as most of the major Mexican figures in decades of transboundary water issues. Also at the conference were representatives from the state users and irrigation districts in the United States, as well as representatives of the U.S. Fish and Wildlife Service, the Bureau of Reclamation, and the Chair of MSCP, Jerry Zimmerman of the Colorado River Board of California.

Until this moment, the presentations had largely been technical and noncontroversial: reports on processes, reports on flows, and reports on research. Everyone realized that Bill's presentation might herald the first step in a dramatic challenge to the law of the river and business as usual.

The moderator from the U.S. Department of the Interior betrayed his uneasiness by making several feeble jokes at Bill's expense. Bill sloughed them off, gathered himself, and launched bluntly into an attack on the MSCP and the Department of the Interior. "Under the law of the river," he said, "wildlife have always come last. While the three states in the Lower Basin struggle to figure out what to do with the MSCP, two of four of the native fish in the Colorado River are extirpated. The willow flycatcher? They still have no handle on what to do for it. But the real problem with the MSCP has been the problem of scope. Defenders of Wildlife made six or seven proposals to include Mexico in the equation. We believe we must include the whole ecosystem in the process. By finding a way to get water to the delta, we know we can solve many environmental problems at once.

"The MSCP rejected all our recommendations. They will not consider the delta in the process."

Bill paused dramatically.

"The MSCP is blowing off Mexico," he declared.

He warned everyone to expect a major announcement in the next week. The audience knew he was referring to a lawsuit on the Colorado River. In one sense such a threat was business as usual for people used to water wars along the border, which are usually structured by battles over rights to the Colorado River itself. Yet everyone recognized clearly that this constituted a new stage in environmental concern for the Colorado River and its delta. After nearly a century of being ignored and forgotten and mollified, the environment and the delta had stepped forward and demanded to be taken seriously.

Less than a month later, on December 14, 1999, David Hogan and Bill Snape, along with another lawyer from Defenders of Wildlife, Kara Gillon, delivered a "Sixty-Day Letter of Intent" to the secretary of the interior, the Bureau of Reclamation, and six other governmental institutions charged with managing the river. The notice documented the violations of the Endangered Species Act by the MSCP and other processes. It announced the intention to sue.

He or she is responsible for endangered species and for Native American rights. The bureau did not even consult with the U.S. Fish and Wildlife Service here in the United States concerning U.S. listed endangered species in Mexico."

The goal is to get the delta inserted into the planning process. The suit is not trying to get water for the delta. "The end game is an amendment to the 1944 treaty between the United States and Mexico that includes ecological factors," Bill says, echoing David Hogan's similar comment.

*A fire over the marsh makes another glowing and unforgettable sunset over the delta.*

Right after Bill made his announcement, the chair of the MSCP steering committee rose to speak. Jerry Zimmerman works for the Colorado River Board of California. He confirmed that the committee believes that its mandate for endangered species work stops at the border. It is the central argument used by states and other water interests to legally fend off the environmental challenge to the river management. For Jerry Zimmerman, the effort to save the delta appears to be a disaster in the offing. The agencies move forward with their planning while the suit is pending. "The ESA and the law of the river have not met yet," said Zimmerman. "A train wreck could happen."

The day after Bill Snape makes his announcement at the conference in Mexicali, a number of environmentalists in attendance take a field trip out into the delta. They visit Ejido Luis Encinas Johnson and then, with several people from the *ejido,* head over to the Ciénega de Santa Clara. Jason Morrison and Michael Cohen of the Pacific Institute, long active in delta recovery efforts, had organized the conference and are on this field trip. A small flotilla of researchers and U.S. environmentalists sally forth in canoes and float out onto the Ciénega.

I join them, as does a photographer and reporter from CNN who is covering the Mexican delta for a story on global water issues—an example of the kind of attention the Mexican delta is generating.

Bill Snape and I share a canoe. The evening is soft and orange. We paddle through the lagoons. Thousands of ducks—redheads, ruddy ducks, widgeons, green-winged and cinnamon teal—rest on the marsh, which is a winter refuge for them and an increasingly important part of the Pacific migratory flyway. A huge flock of snow geese, *ganso nevado,* flies overhead, wintering on the Ciénega.

Something about the Ciénega encourages the conversation itself to become more intimate. "The delta is one of those rare opportunities," Bill says, "it's a historic moment. It's not just a chance to make international law. It's that no one cared about the delta until just a few years ago. We have a chance to find a legal mechanism to really help it."

One of the difficulties with the lawsuit as it is currently constructed is that even if it is successful in court, it may not result in bringing water to the delta. The suit could be successful, in other words, and still not do anything. A court may agree with the interesting application of the law to another nation, though the chances of success are not at all clear. But even success would not guarantee deliveries of water to the delta. It would only force the Bureau of Reclamation to consult with the U.S. Fish and Wildlife Service. It is entirely conceivable, says Robert Glennon from the University of Arizona School of Law, "that the consultation could result in other kinds of mitigation inside the United States."

The lawsuit has its merits. Yet most people, including even Bill Snape, believe it can only be part of a larger strategy. Robert Glennon says that litigation itself is not likely to be the answer to achieving international water transfers, which are the key to saving the delta. Robert Glennon and Peter Culp, in a new and very lengthy article, state their belief that there is an excellent opportunity for the administration of George W. Bush to work with the president's close friend, Mexican President Vicente Fox, to effect water transfers.

While the lawsuit is pending in federal district court in Washington, D.C., Bill Snape agrees that the key in all of this, ultimately, is to locate specific sources of water for the Mexican delta. "Of course Mexico loves the idea of getting more water from the United States," Bill says. "Many think the river was robbed from them. But we're not busting our butts over the delta so bigger alfalfa crops can be grown."

Perhaps the most hopeful development in effecting these water transfers lies in the diplomatic efforts to achieve a new ecological amendment to the 1944 water treaty. For both the lawsuit and the amendment efforts to the treaty, what is crucial for a binational approach to be successful is that Mexico step forward and guarantee that any additional water that ultimately might be delivered to Mexico will actually be used for ecological purposes. Efforts on that front are under way.

The Ciénega symbolizes the environmentalists' efforts on behalf of the delta, so rich is it as a habitat. As we canoe, we can hear some of the thousands of Yuma clapper rails, singing from hidden perches. None come out, but they croak all around us, invisible presences.

While we canoe, I also count at least five northern harriers hunting over the marsh. These birds were once called "marsh hawks," which gives an idea of their preferred habitat. They have long tails and white patches on their rumps, and they cruise low over the marsh as they hunt on broad wings. They are one of the "sensitive species" designated by the MSCP, a focus of recovery in the United States, but they are ignored by the process in Mexico. In the winter, they congregate in the marshlands of the Ciénega, as well as in other wetlands of the Mexican delta.

Most exciting to see, perhaps, are the least bitterns, another "sensitive species" in the MSCP process in the United States. These are small herons, rarely seen. I spot one just at the base of the brown reeds. They are chunky little herons with mossy green backs and a rich brown that streaks the chest, perfect camouflage for blending beautifully with the brown tules.

Time slows on the Ciénega with the unmusical chorus of birds, the fierce sky melting into smoldering sunset. The moment turns sweet. The Ciénega is suddenly eloquent in its silence. All this water comes from the United States. Its supply depends entirely upon the United States. Somehow, that evening the political battles and legal maneuvering fade away with the lowering sun.

The marsh and its creatures become their own beautiful defense, their own best argument for finding legal guarantees that water supplies continue.

One of the most successful of the *campos*, or ecotourism resorts, along the Río Hardy is located just above the levee on the river. A small lake has formed here that continues to cater to American tourists. Campo Mosquedo, as this place is called, is located up river from the Cucapá village El Mayor, and Campo Muñoz of German Muñoz. Along these spring-fed waters Jesús Mosqueda and his son Javier run a successful and busy business by the lake. They have a restaurant and several cabins.

The reason Jesús has succeeded in business is because he took on the U.S. government after horrible floods in the delta in 1983. He defied the odds and the enormous resources of the United States and the Bureau of Reclamation. His defiance not only led to his unique prosperity on the river. It made him a folk hero among Mexicans in the delta. Most Mexicans involved in water politics in the Mexicali valley know of Jesús Mosqueda and his lawsuit against the United States.

Bill Snape is eager to meet Jesús and Javier. Javier is the leader of the ecological association of river users, abbreviated AEURHYC, and is thus one of the co-plaintiffs with Defenders of Wildlife in the lawsuit under the Endangered Species Act. Bill also wants to talk with Jesús because Jesús's earlier case is a legal precedent. That case forced the U.S. government to acknowledge that it knew floods would devastate the delta, which means that the United States acknowledged responsibility for the damage done by the floods—an illegal "taking" of property without compensation—beyond the U.S. borders.

Jesús is an unlikely hero. He is wearing a white cowboy hat and speaking on a cell phone when we meet him. He has kind eyes and a gentle manner. Bill and I drive up to Campo Mosqueda and find him, along with Javier, near a large pond of freshwater the Mosquedas have acquired. For some time now they have been experimenting with a new business in the delta—freshwater shrimp farming.

Today they are harvesting their first batch of shrimp. When we arrive, several men from the regional fisheries department are up to their hips in water, catching shrimp in nets and dropping them into plastic buckets. To Jesús and Javier, this is an experiment in developing a new and sustainable economic base in the delta. They estimate they have 250,000 shrimp now in the pond.

When we ask him about his lawsuit, his face lights up. He clearly enjoys telling the story. The whole affair, he says, was a long drawn-out mess that took years to resolve. It was, he says with a wink, *un chicanazo*—local Mexican slang that means something very weird that somehow manages to turn out well beyond all dreams.

The floods in the delta began in 1979 and lasted for years. The really big flood, though, came in 1983. During this time the Bureau of Reclamation released about two and a half times the treaty water usually released per year since the Glen Canyon Dam filled Lake Powell. Before the first flood, Jesús says he was in Modesto, California, visiting relatives. He came back to find his whole resort under water. Then came the 1983 flood.

"The whole resort was flooded."

"Under two feet of water," he says. "And all the land surrounding it. We piled eight thousand sandbags for a kilometer up the river. We lined plastic everywhere. It was pure mud everywhere." *Puro lodo.*

Many people went out of business. Jesús almost did as well. He had moved his family to the delta in 1958 to open a small farm. He says he did not want to shut down.

"I started selling tacos from a stand. Then I opened a cantina in

1971. I brought my family here because I love the river. I love the lake. I was not going to shut down," he says.

After the flood, he talked to an American friend. The American told him that in the United States, when someone takes away your rights, you can sue. Jesús called a lawyer in Phoenix. The lawyer told Jesús that a complaint against the U.S. government had merit. But it would cost a huge amount of money. And it would take a long time.

Jesús decided to go forward. The Phoenix lawyer referred him to an attorney in Los Angeles named LeRoy Abelson. Soon after the 1983 floods, Jesús filed suit against the U.S. government for illegally destroying his property, without giving fair compensation. The case was called *Gasser v. United States*.

"Jessie was a great client," LeRoy says now, years after the event. "The U.S. government brought in all these big world experts on deltas and flooding. They took a really arrogant attitude. Jessie was a terrific witness. He gave direct answers, he had a sense of humor, he was unflustered, he knew what he was talking about.

*Jesús Mosqueda González successfully sued the U.S. government for damages after the devastating floods of the 1980s, making him a local hero in the Mexican delta.*

195

"It was a fascinating case. I had to go clear back into the Pleistocene to deal with tectonic uplift. It was really interesting."

The case was over what is called a "taking" under "eminent domain" in the Fifth Amendment to the U.S. Constitution. The government is not allowed to "take" property for public purposes without compensating the owner. Jesús's lawsuit alleged that the U.S. government "took" his property, specifically his "easement for a flood," without compensation.

Two issues were at stake. One was whether the U.S. government was responsible for paying a non-U.S. citizen. LeRoy says he found cases as far back as the Revolutionary War for precedents: "Lots of cases, in the Philippines, in Micronesia, lots of cases, establish that a foreigner could claim damages."

Also, the suit had to prove that the regulation of the Colorado River caused the floods and the damage in the delta. "But for the river regulation, these floods would not have happened," says LeRoy. "We didn't have to prove the government intended to flood the delta, or meant to do it. Just that they destroyed property without compensating."

In establishing this, LeRoy says, he used aerial photos, satellite photos, and Godfrey Sykes's book, *The Colorado Delta*, to show what the delta had been like before the dams. LeRoy and Jesús were able to establish that flood patterns had been fundamentally altered by U.S. regulation of the river, dating as far back as 1905.

Jesús says that the U.S. government tried to claim that Mexicans were better off because of the dams. This kind of paternalism makes him angry even as he remembers it. Somehow, Americans had been doing him a favor by damming the river and diverting the water.

"I said, 'No!' I said that we had no benefits from the dams," Jesús says. "We had more freshwater—*agua dulce*—here before Hoover Dam. In 1910, Mexicans were selling water to the Americans. Then the dams go in and we get only recycled water. We get water full of salt. The salt goes way up after the dams. It's too salty."

"We lost our water because of those dams," he concludes.

Jesús says the U.S. experts tried to claim the delta was always flooded before the dams, that the dams brought flood control. "We brought in an old Cucapá woman to testify. She knew the area. She said in court that it was not flooded here. It was mesquite forest. Beautiful. We raised watermelons and sweet potatoes."

LeRoy is not only a lawyer. He is also a civil engineer. He says that what they proved in court was that the changes caused by the dams and diversions built by Americans fundamentally altered the patterns of the river and its flooding in the delta. It was a *caso raro*, according to Jesús, a rare thing. The judge called it a "great scandal."

The case took five years, start to finish. In 1988, the judge ruled in favor of Jesús. After the judgment, the Mexican government even began to pressure

Jesús. The Mexican government sued him on complicated grounds for the money he had been awarded. Jesús believes that the Mexican government itself got into the act because it was afraid this border incident would jeopardize delicate relations with the United States. The Mexican president got involved. LeRoy says he even had a meeting scheduled with then-President Zedillo.

"But the governor of Baja California backed me," says Jesús.

"I believe Jesús's view of this pressure is correct," says LeRoy. "I even had a meeting with the President of Mexico. Zedillo himself. Over this. It was all a really big dispute. It became Baja California versus Mexico City. Finally Mexico dropped it."

It took considerable courage for one man to defy the both the U.S. and Mexican governments. Bill Snape views this case as a precedent for the endangered species lawsuit against the Bureau of Reclamation, as well. It establishes that the U.S. government is responsible for its actions in the delta, according to Bill.

LeRoy agrees. "My belief is that the people in Baja are being ignored by the United States. And the United States is risking further lawsuits."

After we catch shrimp, we go back to the restaurant at Campo Mosqueda. We sit outdoors under the ramada, amid the pool tables and the bougainvillea. A hummingbird darts among the bright red blossoms. A kingfisher cackles over the placid water. Jesús and Javier cook the shrimp for lunch. We share a lovely meal along the river.

I ask Jesús how much he got in his judgment. He smiles coyly. He won't say. I know it was for an amount in six figures, staggering for a man who began selling tacos in the delta. He says that he was written up in *Ripley's Believe It or Not.*

"I must have been very angry to buck the U.S. government," he says. "But I didn't really want the cheese. I just wanted out of the trap."

He looks around. Then, referring to the lawsuit, he signals the hope he and others in the delta feel about the endangered species lawsuit. It is an attempt to get justice after a century of watching the delta itself destroyed by the dams.

"For that suit," he says, "I have prospered."

Whether it is likely that a lawsuit will in itself achieve the goals of additional water for the Mexican delta is not clear. Certainly for a nonlawyer like me, the legal issues grow complex and murky. What is clear, though, is that if there is a will to find a solution to the delta, there will be a way. We need to build a national consensus that we have an obligation to consider the effects of the dams on the delta below the border. That is perhaps why I admire Jesús as he tells his story: not for the lawsuit, per se, but because he stood up for what he believed are his rights. His son Javier is following the same path in organizing the users of the Río Hardy region to work for more and cleaner water.

Out of their dream for a more prosperous delta, they hope a way can be found to reach across borders.

# TEN

# Muddy Waters

ENDANGERED FISH: Totoaba
ENDANGERED MARINE MAMMAL: Vaquita
COMMERCIAL SPECIES: Blue shrimp
PLACE: Delta estuary at the head of the Gulf of California

VERMILION SKY OVER vermilion sea. I am sitting on a lonely and largely abandoned pier, waiting for a scientist who is coming to pick me up in a boat. I am not here for the sunrise. I am here to learn about the once-astonishing fish and shrimp productivity that existed in the shallows of the Gulf of California, at the mouth of the Colorado.

But I am carried away, waiting in the near dark, by a sunrise that is almost agonizingly beautiful. The incomparable sunrises and sunsets are among the desert delta's most superb features.

It's not that such skies are frequent. They're not. Normally, in fact, the delta sky is a cloudless and uncompromising blue. But when the clouds move into the delta basin between the Baja Mountains and the mesa over the Sonora Desert, coming across the Baja, they act upon the light and air in such a way as to create some of the most spectacular sunrises or sunsets anywhere, filling the entire sky. The perfect condition occurs in a broken cover of clouds, or when there is a break between horizon and clouds. The sun breaks through and lights the bottom of the cirrus and cumulus clouds.

This morning the sky is a glut of colors—incandescent yellows near the sun, orange like embers, pinks and crimsons that fade into purple and lilac toward the edges. The sky is so brilliantly red and violet that the sands themselves have taken on a similar the cast. Across the sands, a recent coyote has left its tracks on

◄ A vermilion sunrise over the Vermilion Sea. The Santa Clara Slough is an arm of the Sea of Cortés up the east side of the delta.

▲ Marbled godwits are common shore-birds on the intertidal mudflats of the delta.

a rippled dune. It looks almost as if the coyote had walked across the dune and disappeared straight into the flaming sky.

By the time Salvador Galindo-Bect arrives, I am so swept up in the morning that I can hardly contain myself. He gathers up some gear while I photograph. Shorebirds scurry near the water's edge, little more than dark silhouettes against the mud. They use their wonderful and extravagantly shaped beaks to probe for food. Their names are evocative in their own right, a kind of avian poetry, many of them sounding like the birds' actual calls—long-billed curlews, marbled godwits, whimbrels, and straight-beaked willets, along with the much-beloved avocets and long-necked stilts.

These birds are signs of the fertility at the mouth of the Colorado River. Salvador, whose nickname is Chava, has been working for years to document and prove that richness. Chava is a researcher at Universidad Autónoma de Baja California in Ensenada. His research has proved that the waters here at the head of the Upper Gulf serve a unique and vital function in the sea life of the Gulf of California, and, just as important, for the fishing economies of the Gulf that depend on this sea life.

Together with botanist Ed Glenn, Chava has demonstrated that the delta's estuary depends on the freshwater from the river to fulfill its role as nursery to the sea. Shrimp harvests crash when the river quits flowing. There have been bumper crops after the big flood releases, what he calls a *respuesta biologica,* or biological response, to the freshwater. Chava is now working out the explanation for how this *respuesta* works.

"The shallow waters around the mouth of the Colorado River are the great nursery of the Gulf of California," he says as I climb aboard the small outboard boat. "It's a nursery—*area de la crianza*—for many fish and shrimp species. And it's all stimulated by the waters of the river."

He has picked me up at the Slough of Santa Clara, a saltwater arm of the sea that reaches up into the fertile mudflats of the delta toward the Ciénega de Santa Clara. Chava and I putter out of the slough and toward the sea. But the sea does not open up here into a broad, flat sheet of water. To the west is a low flat island, Isla Montague, and behind it is another, smaller mud island, Isla Pelicano. These two islands are like little appetizers sitting almost in the mouth of the river. The river hits the sea, and then sweeps around the islands in broad channels, before joining the open sea.

These are muddy waters here at the river's mouth. Historically, the load of fine alluvial silt carried by the river, supplied through geological time, made the waters at the head of the Gulf muddy and very rich. "It's like a very fertile lagoon," says Chava. "It's fabulously rich, amazingly fertile."

Now it is the political murkiness upriver that makes these estuary waters so muddy. The Gulf of California is essentially an inland sea. It has developed its

own fish, its own internal dynamics both in waters and in the creatures that have, over evolutionary time, adapted to these waters. The Gulf is famous for its productivity in fish—586 different species are recorded for the Gulf as a whole. Of these, 92 are endemic species—they evolved in the isolation of this sea, and exist nowhere else in the world. This rate of endemism is very high. About 50 of the species of the Gulf, including at least one very important endangered species, have their principal distribution in the northern or Upper Gulf.

The shallow waters and the enclosed nature of the sea create the famously high tides and strong currents of the Upper Gulf. The result is a high level of turbity in the waters—muddy waters. Combine that with the freshwater coming into the sea from the Colorado River—the only freshwater source for these waters—and you can grasp its startling richness.

Chava is putting hydrology together with biology—water flows and life cycles—to explain the unique role of this part of the Gulf in the life histories of several species of fish and shrimp. The situation here is the same as it has been virtually everywhere else in the delta. This place where the river meets the sea, the estuary of the Colorado River, had been almost entirely ignored by scientific researchers. It was certainly ignored when negotiators signed the 1944 United States–Mexico Water Treaty. The result was that no one had any idea what choking off the river's freshwater flows into the sea would do to the fisheries

*Salvador Galindo-Bect, "Chava," steers his research boat near Isla Montague in the background. His research helped demonstrate the link between the crash in shrimp production in the Gulf of California and the dams upriver.*

here. As Chava puts it, "My research is real proof that the lack of freshwater has a huge impact [*tan impactante*] on this nursery to the sea."

Along with geologist Karl Flessa's work on clams in the delta, Chava's work has demonstrated a startling and compelling connection between the several endangered species and the several fisheries in crisis. In the United States it is customary to view nature and culture as opposed, ecological health versus economic prosperity. Chava's research suggests the two are not opposed. In the delta, when it comes to fish, ecology and economics are inextricably connected.

Together with Ed Glenn, Chava conducted research that showed the close link between the ecological health of this part of the delta and fishing success in the Upper Gulf. The researchers documented what local fishermen have been saying for years. For decades the fisheries from Puerto Peñasco, El Golfo, and San Felipe—the three main fishing communities in the Upper Gulf—have been in decline. The downward trend came to a head in the early 1990s, when the fishery went into crisis. Then San Felipe, on the other side of the Gulf, had about 15,000 people. El Golfo de Santa Clara, just south of the slough here, is much smaller. Only about 600 people lived in this little sandy-street village on the Gulf. The fishermen in both communities used to operate as cooperatives. In 1990, San Felipe had about 40 large fishing boats working out of its port. El Golfo had 15. By 1993, that number had dropped to 23 and 8, respectively, and the villages' populations dropped correspondingly.

The crisis was caused by a crash in the fishery for blue shrimp, *Penaeus stylirostris*. This was only the latest of the crises in the fishery here. The most important commercial fish in the Upper Gulf for decades was the totoaba, *Totoaba macdonaldi*, known also as the Mexican giant sea bass. It was huge, some two meters in length and weighing up to 135 kilograms. From the 1920s to the 1940s, the totoaba became the principle commercial species of these waters, with fishermen supplying the huge U.S. market for the totoaba. In 1942, the annual catch was 2,261 metric tons. By 1975 the catch had fallen to 58 metric tons.

The decline was so precipitous that scientists began predicting that the totoaba would go extinct. In 1975, all fishing for the totoaba was banned. In 1976, the totoaba was listed as threatened under the Convention on International Trade in Endangered Species—a move to protect the species by cutting off imports. In 1979, the United States declared the totoaba an endangered species—a move to cut demand in the United States and to protect the species from poaching.

For as long as the fisheries have been in decline, the fishermen have been blamed for the problems—overfishing the resource. There is truth to that claim. These poor communities have put enormous pressure on the fish and shrimp. When the Reserva de la Biósfera Alto Golfo de California y Delta del

Río Colorado was created—one of the largest biosphere reserves in the world—huge portions of the Upper Gulf were included expressly to protect the endangered totoaba.

For both the blue shrimp and the totoaba, local fishermen bore a huge load of blame. Fishermen howled loudly at not being allowed to fish. When the number of boats declined in the villages in the 1990s, for example, pressure on the two species actually increased, even as catches decreased. More fishermen took to shrimping in their small boats, or *pangas*. Totoaba were caught in the nets. Plus, poaching of adult totoaba in this remote region was a common practice into the early 1990s.

Mexico has made moves to manage the fisheries resources in the area, declaring the totoaba endangered and creating the biosphere reserve. The country's environmental sensibility is still in the early stages, but efforts are being made. Still, the fishery here demonstrates one of the great difficulties in Mexican environmental management: enforcement. Poaching has been and remains a major problem. Mexico is trying to grapple with enforcement issues, but does not yet have the situation under control.

Even as Chava and I are talking about the crash in the fisheries on the estuary, we watch at least one hundred boats of local fishermen breaking the law—poaching fish. It is the first day of the corvina season. While fishermen are allowed to catch fish in the biosphere reserve's buffer zone, fishing in the reserve's nuclear zone is prohibited. But we watch boat after boat stream past us, heading up into the river, into the prohibited area. Some boats are in the buffer zone, catching corvina, hauling them into their boats.

A guard with the biosphere reserve told me that in one season, fishermen made fifteen threats on his life. They are like a mafia, he claimed.

"No fishing whatsoever is permitted here," Chava says to me. "We're inside the nuclear zone. But it [fishing] exists."

The same phenomenon exists at the fishing camp for corvina along the Colorado River that I visited with Don Onesimo and Pedro Bueno Bueno.

The biosphere reserve has only one guard to try to enforce the laws. It is an impossible task, and the local fishermen can be very aggressive. They are poor. Fishing is their only income. At meetings, they scream and yell. The guard told me that the government is making progress with enforcement and education. It's fair to say that these fishermen have only begun to learn about long-term environmental and sustainability issues. As the guard told me, "Over time things will get better."

Enforcement has to improve in Mexico. Yet fishermen insist that the problem does not entirely rest with them. The fishermen say, for example, that most of the problem relates to the amount of water coming down the river. Whenever the river flows high, there will be good fishing for two years following.

*Fishermen from El Golfo de Santa Clara, a fishing port on the east side of the Gulf of California, haul illegal corvina inside the biosphere reserve. Illegal fishing by economically depressed people has contributed to the difficulties of environmental protection in the delta.*

Chava and Ed Glenn decided to check out what the fishermen were saying. That led to Chava's current research program—to see whether hydrological conditions are related to fisheries productivity.

❧

"Ed Glenn and I did not believe it was only overfishing," Chava explains as we head up toward the very mouth of the Colorado River. "There was a virus too. But we believe even more that a change in the hydrographic characteristics in the area of reproduction, this nursery, was a crucial component."

Simply stated, Chava and Ed wanted to see if the loss of freshwater flows from the Colorado River played a role in the crash of the shrimp and totoaba species and fisheries. The study was the first stage of proving that this *area de la crianza* is crucial to the fisheries of the Gulf and is dependent on freshwater flows from the river. Chava and Ed compared shrimp "landings," or catches, with the flows of the river. They looked particularly at what happened to the shrimp catches in El Golfo and San Felipe in the years of the high El Niño flood flows. What they got was a sharp picture. During the years of the El Niño flood releases from the dams in the Colorado River, enough freshwater reached the Gulf to simulate pre-dam conditions. There was a significant correlation between the high flows in those years and greater shrimp catches.

As interesting, Chava found that the correlation between increased flows and increased shrimp catches was very high—much higher—in the year *following* the flood releases, in the next year's catch. This finding was almost exactly what the fishermen's local wisdom had claimed. Harvests peaked sharply.

In the boat, Chava stops the motor and we drift while he roots through a muddy briefcase he has had on the bottom of the boat. He is wearing a dark fleece jacket for the cool weather and a bright white hat. He has been working hard for several days on the water. His face looks tired. He has bags beneath his eyes. But there is something soft and *simpatico* about his face, about his earnestness as he tries to find a graph. He pulls it out. He shows me the decreasing flows to the Gulf over the last century. He shows me graphs of crashing fisheries.

Then he shows me the El Niño years—and the higher catches of shrimp following those high levels of freshwater flows. "El Niño proved it," he declares. Behind him I notice the clouds sweep dramatically across the sky. "There's a biological response to the impulses of freshwater." The water samples we are taking this morning are part of a research program Chava is conducting to try to understand the mechanisms of this biological response to the freshwater inputs into the sea.

The answer is beginning to emerge. Chava explains the way the ecosystem works in this nursery. Over nearly the last seventy years, the freshwater flows have been about 1 percent of what they were before the dams went in upriver. In an earlier chapter, I described how that has made the delta into an "anti-delta," characterized by erosion rather than accretion. This lack of freshwater and silt affects the water characteristics and currents and as well. Where the Colorado River meets the sea, it is no longer an estuary.

Chava calls it an "anti-estuary." By that he means that in a normal estuary, freshwater mixes with the salt water of the sea. The salinity therefore decreases as you go up river. Not in the Colorado River. Salinity actually increases from the sea into the river. The increase in salts is not from pollution in the river. It is from the lack of water and the increased evaporation, which in turn concentrates the salts. "The decrease of Colorado River freshwater has drastically changed the ecological conditions. [The river] is now an area of the highest salinities in the whole Gulf."

In the pre-dam estuary of the Colorado, the turbid waters here were high in nutrients carried by the river and stirred up by the huge tides. But the dynamics of this nursery are more complex and more interesting than that. In the post-dam estuary, there are still nutrients in the waters, stirred up from the mud. But two other factors have changed significantly: the salinity and the currents in the Upper Gulf. In the natural nursery in the "lagoon" of the sea at the mouth of the river, the lower salinity protected young shrimp and young fish from big pelagic or seagoing predators. These predators would not enter the less-saline waters pouring in from the river—and hence the nursery served as a kind of refuge for fish and shrimp at vulnerable stages of their lives.

The blue shrimp come to this nursery at the postlarval stage of their lives—a time when they are young and particularly vulnerable. They find both food and protection in these waters. They would normally come to this nursery at exactly this time of year, when Chava and I are collecting water samples, in early April. They would come on sea currents from the deeper waters to the coast, and then ride currents along the shores up to the head of the Gulf.

"Without sufficient water, the currents have changed," Chava says. The saline waters now have vertical currents, as water moves up and down, which are stronger than the older counterclockwise currents that would bring the post-larval shrimp into the area. The shrimp now cannot reach their natural refuge.

"The larvae die in the sea and never arrive at the coast," says Chava.

"So what we're trying to demonstrate now is that, without freshwater flows, there is no longer a nursery, and so that's why there's no fishery," he concludes.

Chava reminds me of many Mexican people I have met in the delta. He is diffident when he begins to speak. But his passion for the topic enflames him and he has grown animated through the course of the morning. He talked with me for almost four hours nonstop about the Upper Gulf, its lack of freshwater, and the crash of its many species of fish and crustaceans.

He has proved quite clearly that there is a correlation between shrimp productivity and freshwater—and the economic devastation that came with the loss of the freshwater. He believes as well that totoaba, one of the premier endangered species of Mexico, may also have been devastated by the loss of water. It is well documented, for example, that totoaba migrated to the Upper Gulf at a key stage in their early development. One study in the 1980s showed that 73 percent of the juvenile population was found in the waters near San Felipe; 20 percent was found near the mouth of the Colorado River near El Golfo.

One of the most highly regarded experts on the Gulf of California is Saúl Alvarez-Borrego, a researcher in the Division of Oceanography at the Centro de Investigación Científica y de Educación Superior de Ensenada, in Baja California (CICESE). Alvarez-Borrego has worked with Chava on his research on the nutrients of the Upper Gulf, and he confirms the lack of freshwater, along with the capture of juvenile fish in shrimp nets, are two reasons the fishery no longer exists.

Karl Flessa's proposed study of the isotopes in the otoliths of totoaba, the bony structure in the fish's inner ear that records freshwater oxygen values, can help determine exactly what the role of the freshwater is in the life cycle of the totoaba. It is expected to be important. The totoaba is one of the really significant endangered species in the delta of the Colorado River. If the lack of water is definitively connected to its endangerment, this species will demonstrate

even more strongly the nexus between endangered species, endangered fisheries, and the dams on the Colorado River.

This morning we do not go all the way into the mouth of the river. Chava had collected the samples he needs from there the night before. Instead, we turn around near the northern end of the *isla* and head back to the port of El Golfo. On the way back, we are joined by a pod of bottlenose dolphins. They swim in our bow wake, dive beneath the boat, even leap in lovely arcs from the blue water of the open Gulf. Behind them lie the desert bluffs—the ancient bed of the river.

I keep a vigil for the extremely rare vaquita, *Phocoena sinus*, the "little cow," as its called—the world's smallest and rarest marine mammal. It is almost never sighted. About once or twice a year, a fisherman pulls up a vaquita that has gotten entangled in the fishing nets. That's about the only proof that this deeply endangered marine mammal still survives in these waters. The role of freshwater in the life of this little dolphin is less clear, as is the role of the dams in its decline. What is clear is that the gill nets used by fishermen in the area killed thirty-nine of these dolphins between 1993 and 1995, almost one-fifth of the

*Bottle-nosed dolphins romp near shore in the upper Gulf of California. Another porpoise, the vaquita, is the smallest and rarest marine mammal in the world—and is found only here. Note the Sonoran Mesa in the background.*

207

total remaining estimated population of two hundred. Scientists and conservationists are calling for a complete ban on gill nets in the biosphere reserve, and for extending the reserve boundaries to cover the entire range of the vaquita in the Upper Gulf.

Despite what we do not know about the vaquita, we do know is that this species was only discovered in the 1950s. It is feared that it may be extinct within just a few years. Only half a century since being discovered, it's nearly gone.

For some time, we cavort with the dolphins. Chava drives in circles while the dolphins jump and dive and eye us from the waters.

We have a lot to account for in the way we have treated the estuarine ecosystem at the very mouth of the Colorado River. It is abused by a lack of water. It is abused by overfishing. There is plenty of blame to go around in this. Both Americans and Mexicans can be called to account. "It's an ecosystem that we don't know how to manage yet" says Chava. "The Mexican government needs to fund lots more research. Now there is no good relationship between management, the ecosystem, and social problems."

Yet for Chava the deeper and more compelling implications of his research lead him to the root of the problems—the lack of freshwater. And that takes him to the 1944 United States–Mexico Water Treaty. The fish, the shrimp, the dolphins—none of them was considered when the treaty was signed. No one examined the impact of the law of the river on the nursery here in the Upper Gulf—one of the most productive and fertile fish and shrimp nurseries in the world. Instead, the water was simply taken away.

"That's the principal problem with the international treaty of waters," Chava says bluntly. "When Mexicans and Americans signed that treaty, that gave a quota of only 10 percent of the river to Mexico, they never considered the ecological impact of the restriction of freshwater to the ecosystem."

It is an ecosystem heavily stressed at both ends, in the United States and Mexico not enough water to produce fish and shrimp, too much pressure from fishermen. The law of the river recognizes none of these conditions. Yet it controls the basic terms of existence for life in the area, human and animal, fishermen and fish. The crisis offers eloquent testimony to the need for cooperative efforts between the two countries—and for a revision of the treaty that fostered this circumstance. Having claimed for itself the lion's share of the water, the United States bears considerable responsibility for this ecological devastation—both cause and remedy.

The Americans dammed the river. The enterprise to control and harness the Colorado was entirely American—for better and for worse. The economic development of this part of both Baja and Sonora has consistently followed in the wake of the engineering developments on the river by the Americans.

Chava speaks forcefully, indignantly, about the treaty. "If the United States wants to maintain an image of a country that cares about ecological problems throughout the world, that's vigilant," Chava says, "then it needs to care not only about problems generated by other countries, like Mexico, but also about ecological problems that it generates itself."

The dolphins dive as they swim off toward the desert shores. Their flukes rise above the waves. The outboard motor whines as we head to the landing beach at El Golfo. The light is magic—soft and subtle, a pewter blue beneath the wash of clouds in the sky. Some of the clouds are sweeping and white, others clotted and gray. The light falls in a faint veil of blue over the sea and island.

*Brown pelicans fish at the port of El Golfo, in the Ciénega and throughout the delta.*

209

# ELEVEN

# The Delta and the New Western Water Wars

SPECIES: Burrowing owl
PLACES: Mission Inn, Riverside, California, and agricultural levee, Mexican delta

ON SEPTEMBER 29, 2000, the then Acting Deputy Secretary of the Department of the Interior, David Hayes, addressed a symposium meeting on the Mexican delta of the Colorado River. The symposium was held at the swanky Mission Inn, in fast-growing Riverside, California. The inn has a gracious and relaxed Mexican feel to it, built in the style of Spanish missions in the area. It's the kind of place where presidents of the United States stay, and a picture of Franklin Roosevelt is displayed prominently in the lobby, commemorating one of his visits.

That the Mexican delta would command such a posh venue, and such a large crowd, signifies the changing political geography on the lower Colorado River. The delta is coming into its own, and there was a palpable sense of hopefulness in the crowd. It must have seemed to everyone that a kind of sea change was happening: The delta had perhaps finally arrived as a major issue inside the United States.

There were bilingual interpreters and speeches and presentations in both English and Spanish. In the audience was anyone who has ever done a stitch of work in the Mexican delta—lawyers, academics, scientists, water users, activists and organizers, journalists.

José Campoy, the director of the biosphere reserve, was there. He gave a speech and slide show on the natural history of the delta.

Botanist Ed Glenn spoke about the delta's biological recovery. He reminded everyone that the delta needs about 1 percent of the Colorado River flow to sustain its recovery.

◄ *Looking west from the eastern Sonoran Mesa, the lower delta and its river glows in soft evening pastels.*

▲ *Burrowing owls have prospered with the growth of agriculture in the delta. Mexicans call them "agachones," the squatter, for their habit of bobbing and crouching.*

Carlos Valdés, one of the first to advocate for the delta, was a center of attention. Only a decade ago, he had begun his research when there was no attention given to the ecological concerns in the delta. Carlos was busy giving interviews to local newspapers.

Javier Mosqueda, from Campo Mosqueda, spoke on behalf of the Río Hardy and the users association in the delta.

Juan Butrón was there from the Ciénega de Santa Clara, listening.

The organizers had given small grants to many Mexicans from the delta so they could afford to make the trip and stay in this classy hotel. I wondered how Juan felt, comparing his small *ejido* with the wealth of this place, this gathering. The symposium had a bilingual title, ". . . *hacia el Mar de Cortés*"; ". . . to the Sea of Cortés." It was a working symposium on the Mexican portion of the Colorado delta and its relationship to United States water politics. It was also a "who's who" gathering of people working on the delta.

The main meeting room of the Mission Inn was full to overflowing with Americans and Mexicans. There must have been 200, maybe 250, people there when David Hayes spoke in the afternoon. His speech left a number of people unsatisfied, as he carefully threaded his way through the tangles and pitfalls of vested interests and competing claims concerning water in the West. He spoke in generalities, careful not to commit the United States to anything.

But the significance of his address was less in its content than in his willingness to regard the delta seriously. The Mexican delta may not be popular with the states. But at least the federal government and the Department of the Interior had recognized it.

"At first blush many on the U.S. side of the border are inclined to say the delta is not an issue. Not our problem," Hayes said. "Solutions foisted on us by Mexico or environmentalists would clash with key interests." Yet the federal government, he argued, was not willing to let it die there.

"The Colorado River is an international river," he declared. "We have a binational working relationship with the government of Mexico. If this is an issue to Mexico and to American groups, then we cannot assume all hell will break lose if we open this door. It is appropriate now that we join with Mexico and start talking about the Colorado delta—without assumptions about the nature of the problem and its solutions. We need to enter into collaborative discussions in good faith. The first step is to learn about the delta."

That he was willing to open up the topic was groundbreaking. The delta had been growing as an issue in the United States for some time, increasingly a darling of environmentalists who recognize the natural history values of the wild delta in Mexico, as well as the social justice issues. And that plops the Mexican delta of the Colorado River squarely in the middle of the new water wars in the West.

"Change is here concerning the delta," Hayes declared. With that pronouncement, it was almost as if the delta had been declared a major national environmental issue. For all the advocates of the delta in the room, that really means one thing: nothing less than an amendment to the 1944 water treaty between the United States and Mexico.

To understand the significance of getting an amendment to the 1944 water treaty, and the Herculean effort that it represents on behalf of the delta, you need to understand some currents in the politics of the new water wars in the West. Because even as hope for the delta is growing, the threats to its water future are gathering all around. Clearly sentiment for and knowledge of the delta in Mexico has grown. The number of people inspired to work on behalf of the delta is swelling. Yet these advocates ought also to be terrified. The pressures on water are growing almost daily in the American West: Politicians and city boosters have dreams of greater growth, and if the major users in the United States could have their way, the delta would not get a drop.

In an Associated Press newspaper article that I read in the Palm Springs *Desert Sun*, Ed Glenn is credited with a searing quotation about U.S. water users: "The environmental argument doesn't carry much weight. Most of these guys wouldn't put a cup of water on a dying plant in Mexico" (Rosenblum 2001).

When I asked him about it, he told me that it's a great quote. The only trouble is, he never said it. But it's the kind of quote *someone* should have said, he added, even if he did not.

The quote hints at the ferocity of feeling arrayed around every drop of water on the river—and the hostility as U.S. water interests line up against giving water to Mexico for the delta. The delta has a formidable array of foes. Water politics along the Colorado River has always been a mess of conflicting interests. But a new round has arrived—a new wave, if you will—in response to the unrelenting pressures of growth.

To understand the intensity, the very civilized fury, in the battle to get an amendment to the 1944 water treaty on behalf of the delta, you must understand where the delta fits into the new water wars. California 4.4 Plan, All-American Canal Lining, Surplus Criteria, the Salton Sea—these are all looming issues competing for the limited water resources of the West and are clarified in the story that follows. For environmentalists, the delta is in a sense the cause that might be able to make headway against all of them, in the process forcing us to consider the deeper ethical and social issues involved in our headlong rush to growth.

The river, as we have noted, has been overallocated from almost the beginning of U.S. aspirations to harness its waters. By the time allocations were finalized in the 1944 water treaty, the projected overuse of the river was already nearly hopeless. In the treaty, Mexico agreed to 1.5 million acre feet (maf), with a possible addition of another 0.2 maf for "beneficial use" in years of high floods. That was on top of the U.S. allocations to the seven basin states of 16 maf. Pile onto this another 2.5 maf for "system losses"—leaks and evaporation and other inefficiencies—and river users are in way over their collective heads at 20.2 maf. The river, remember, only sustains about 13.2 maf of use per year on a five-hundred-year cycle, according to a University of Arizona tree ring study.

In terms of allocation, we are in a systemic river deficit to begin with—to the tune of about 7 maf overdrawn at the riverbank, as one Arizona newspaper put it. Only two things have kept this system afloat over the last half-century: the unexpected El Niño floods, and underconsumption by the Upper Basin states of their 7.5 maf allotment. According to the Bureau of Reclamation, in 2000 the four Upper Basin states were still only consuming about 3.7 maf.

Neither of these circumstances will last. Experts are now predicting drought, not floods, on the river. Of course, no one can know this for sure. And the Upper Basin states are jockeying into position to take deliveries of what is due them in river water. Some say the Upper Basin states can never use their full allocation of water. Others fear that it is only a matter of time before they both claim it and use it. What is clear is that these Upper Basin states definitely do not want to relinquish their legal rights to more water—and the economic possibilities that water makes possible. Because if there is a true liquid currency in the West, it is water. Water means money.

While a new environmental consciousness has carried the river into a new stage in its management, the river is going through incredible growing pains as well. Human demand is near the limits, and rather than letting up, this demand is burgeoning. More water is needed. The looming prospects for the river and its delta are ominous.

The already overcommitted river now supplies water for up to 30 million people in the United States along its watershed. In the Lower Basin, 23 million people use the river, and that population is expected to grow by more than 8 million people by the year 2020. The arid Southwest is the fastest growing region by population in the United States. Southern California alone, according to Acting Deputy Secretary of the Department of the Interior David Hayes, is growing by almost a million people per year.

This relentless population growth along the river is like a monster that must be fed. As the Lower Basin states began to reach and exceed their legally allocated limits, a new round of planning and a new round of skirmishes and fights over water developed. California's allocation for the Colorado River, by law, is 4.4 maf. Most of this water has historically gone to the Imperial Irrigation District around the Salton Sea—agricultural interests. However, about 1.2 maf have gone to the insatiable cities of Los Angeles and San Diego, shipped through one of the wonders of 1930s engineering, the Colorado River Aqueduct.

The growth of southern Californian cities has the Metropolitan Water District of Los Angeles predicting a 2.1 maf shortfall by 2020. Yet for years now, southern California has been grabbing more water than it is entitled to. It has been consuming up to 5.3 maf per year.

Clark County, Nevada, is the home of Las Vegas. It is the single fastest growing county in the United States. Nevada gets only 300,000 acre feet (af) of Colorado River water by law—the Boulder Canyon Act. For this empty desert state, that might have looked like more than enough in 1928. Now it is completely inadequate. Nevada is searching for new sources of water—from groundwater to expensive transfers from other sources. It has applied for the rights to more than 800,000 af of ground- and surface water.

Arizona has just completed its half-century engineering project, the Central Arizona Project, a 336-mile canal that carries Colorado River water to Phoenix and Tucson. Tucson will now be consuming about 1 percent of the river, meaning that Arizona will be able to use its full allocation of the river, 2.8 maf. Arizona is even planning to bank water in the ground in a project called the Arizona Water Bank.

It goes without saying that every state in the Lower Basin is heavily dependent on the Colorado River.

*Members of AEURHYC, the ecological association of Río Hardy water users, meet to plan strategy for the conservation and restoration of the delta. Members include (from right) Javier Mosqueda, Ismael Yanez, and Juan Hernandez. Yamilett Carillo (foreground) of Pronatura, Sonora, facilitates.*

215

In addition, new forces have been asserting their legal rights to river water, most notably the Native American tribes along the river, and the environment under the Endangered Species Act and other federal legislation.

With these growing pressures, then Secretary of the Interior Bruce Babbitt invoked his rights as "watermaster" of the lower Colorado River. The focus of his attention fell on California. Without surplus flows from the Colorado, southern California would be in serious trouble. Since cities were not the highest priority on the list of water users, they would be the first to be cut back. Their allocations would go from the 1.2 maf to about half that. In 1996, Babbitt declared that he was going to put southern California on notice. The problem was that California had long been using more than its legally allocated share of the river. The state is also a powerful economic and political force—it has the sixth largest economy in the world. But with pressures on water in the Lower Basin growing, Babbitt took the state on in what came to be called the California 4.4 Plan.

The state's water users were forced to develop a plan whereby they could live within their legal allotment of water—4.4 maf. Several years of bluff and confrontation, blunder and controversy, led to the plan. California was told it had to come up with a plan to cut back from its 5+ maf consumption levels. The state was forced to figure out where it could come up with the difference, reducing consumption from the river by at least 0.7 maf.

A number of measures were devised. A 200,000-af transfer of water to the San Diego County Water Authority from the Imperial Irrigation District was crucial, and very controversial—as we'll see later concerning the Salton Sea. Water was also gleaned from efficiencies and savings in the system, largely developing and deploying new technologies (canal linings, automated gates, flow-control devices, better metering) to control the flow of water in canals. California managed to reduce its consumption use forecasts, but not to the point where demands can be met in a normal year.

Not 4.4, but a step in the right direction.

In the process, the cities managed to get an *increase* of water. The water going to Los Angeles and San Diego actually rose above 1.2 maf. As Peter Culp writes in his article on the law of the river, "The inescapable consequence is the long-term shift of water from farms to cities." It is an inevitable trend that will continue into the future. He notes, for example, that in 1994, Arizona devoted nearly 2.2 maf of its allotment to the irrigation of cotton and alfalfa—nearly 80 percent. Alfalfa returns only $95 per acre foot of water used. Cotton, about $192. By contrast, a person in Tucson pays nearly $750 per acre foot for residential water (Culp 2000, 5).

If California were to meet the challenge, it could not do so immediately. It insisted on having until 2015 to meet the demand. Babbitt agreed to a "full Colorado River Aqueduct" until then. Plus, California, with the other states behind it, insisted that the Bureau of Reclamation redefine the meaning of "Surplus Criteria" in the river impoundments. Currently, the bureau holds about 60 maf behind its dams. This amounts to a bit more than a four-year flow of the Colorado River, an insurance policy against drought years.

California had long insisted that this surplus was too much. They argued for a lower level in the reservoirs. In other words, they wanted the secretary of the interior to be able to declare that there was a surplus when there was less water in the reservoirs. That meant the states would get access to more water on a regular basis.

Because the reservoirs have historically been kept at capacity, they have had no room to capture the El Niño floods that have taken place since 1980. The upshot? The delta got the excess. The writing was on the wall for the delta. If the states were to get their way on a new definition of Surplus Criteria, no more flood flows would come down the river's channel to the delta. Such a redefinition would be tantamount to declaring a second death for the delta. The floods would all be captured above the border. The delta would once again be hung out to dry.

The bureau issued its draft Environmental Impact Statement on the various alternatives for declaring a surplus in the river. Environmentalists had argued for an option that promised at least some flows to the delta—Michael Cohen of the Pacific Institute was one of its principal authors. The environmental option was not even included as an alternative. The impact statement by the bureau ducked the delta issue. The bureau claimed that all water that passes the boundary into Mexico is Mexico's responsibility, and the bureau has no authority to dictate to Mexico how it uses its water. It is an argument that has been frequently used in border water politics over the delta. Furthermore, the bureau argued that it had no legal discretion to release water for Mexico. Its releases are mandated by law, except in flood years.

For California and the other six states that rely on the Colorado, lower reservoirs were the price the feds would have to pay to put California on a diet. More "usable" water would be put in the system.

A Surplus Criteria plan was approved in 2001.

California and the other states get their surplus water from the lower draw-downs. California also has until 2015 to meet its legal 4.4-maf allotment of Colorado River water. But as the reservoirs sit at lower levels, flood waters can be more easily captured. The only floods likely to reach the delta will occur from the Gila River, near Yuma—the last river below the major dams on the Colorado.

The picture in the short-term is bleak. No more water will get to the delta without some compensatory legal mechanism. Everyone genuinely fears that the delta will slide back in to its desertified state, as it was between 1935 and 1979, when the dams' reservoirs were filling and the river died an ignoble death in the sands of Mexico.

What will happen with the surplus water when, in 2015, California is forced back near its legal limit of 4.4 maf? Surely that would be a good thing? Then all the states will be living within limits and water can again be available for the delta, the thinking goes. But very likely this will not happen. According to Kara Gillon, a lawyer for Defenders of Wildlife, many environmentalists fear that by that time the Upper Basin states will probably be ready to use their full allocations of the river. The United States has its hands on the tap and, from the delta's point of view, may be turning the tap off for good.

Bill Snape of Defenders of Wildlife put the new round of power politics in succinct perspective. "The California 4.4 Plan morphed into the Surplus Criteria," he says. "It gave fifteen years for a soft landing for California. It's technical gobbledy-gook, but the net result is they'll have a more liberal definition of a surplus, so they'll lower Lake Mead. So the spills to the delta are not going to exist. The concepts were essentially written by California. Simple. It'll shut off the water for fifteen years. And no one has a solution to the problem—except our lawsuit on endangered species.

"It makes ya' wanna scream," he says. "They should be terrified in the delta."

Despite the political complexities, David Hayes' speech signaled a growing willingness in the Clinton Adminstration to address the conservation and restoration of the Mexican delta. It was as though a logjam of inaction and paranoia over the delta suddenly broke. In the next several weeks, major developments made the prospects decidedly improve for amending the 1944 United States–Mexico Water Treaty, to secure water for ecological restoration in the Mexican delta.

A meeting the month following the symposium in Riverside, in October 2000, was perhaps the most dramatic. The Department of the Interior convened a meeting with the explicit goal of working directly on possible solutions to the *problematica del delta*, the problem of the delta. The two federal governments of the United States and Mexico had been engaged in diplomatic efforts for some time on the ecological issue of the delta. In May 2000, for example, they had signed a Joint Declaration in which they pledged to cooperate and to redouble efforts to find mechanisms to improve the conservation of habitats and cultural values in

*The only Mexican dam on the Colorado River, Morelos Dam was mandated by the 1944 Water Treaty and is used to receive and divert water deliveries into Mexico.*

219

the delta. It was then considered the most progressive step on behalf of the delta yet taken by the governments at the highest levels. Both the U.S. Department of the Interior and Mexico's Secretaría del Medio Ambiente, Recursos Naturales y Pesca (SEMARNAP, now SEMARNAT) signed it. Still, no one was prepared for what would happen next.

The result of the meeting in Washington, D.C., was a proposal that left delta advocates pinching themselves. The agency proposed that the International Boundary and Water Commission (IBWC) prepare a "conceptual minute" for the treaty—the first step in achieving an amendment to the 1944 water treaty.

Bill Snape was at the workshop. Representatives from both the United States and Mexico were at this workshop. The main strategy of advocates for the delta had long been what is called a "minute," or an amendment, to the 1944 water treaty that would entail deliveries of water from the United States to Mexico for the purposes of environmental restoration. The idea came up that a conceptual minute would get the process moving—the agencies charged with managing the provisions of the treaty would then work out a process leading to more formal minute or amendment.

Snape said his chin hit the floor when the idea came up and was approved. The 1944 water treaty gave the binational IBWC (CILA in Mexico) authority to manage transborder resources between the United States and Mexico. The international water commission is authorized to build and manage waterworks along the Colorado River and Río Grande, to negotiate further agreements regarding international waters, and to settle disputes over treaty interpretations. Its decisions are referred to as "minutes" because they are written into the record of the meetings, into the meeting minutes. These minutes are binding unless objections are received within thirty days, after which they become part of the governing law of the river. Any program for delivery of water across the border to Mexico, for use in a delta restoration program, falls in the IBWC's lap.

The IBWC commissioners from the United States and the CILA commissioners from Mexico met on December 12, 2000, in El Paso, Texas. They were charged by their respective governments with considering a conceptual minute to the treaty of 1944. Their charge, specifically, was to consider "a conceptual framework for cooperation by the United States and Mexico" on the delta's riparian and estuarine habitat (IBWC 2000, 1). The commissioners followed through on their charge that day. Translating their deliberations out of the formal and legal language of the minute, we discover that they produced a document that outlines a process for determining how water might be found for ecological restoration of the Mexican delta. It does not mandate transfers of water or set up mechanisms for implementing such transfers. Rather it is a major preliminary step, defining a process of joint studies and recommendations concerning water and ecology in the Mexican delta. The commissioners noted that technical water studies are needed; they noted that the integrity of each country's laws regarding the environment must be respected; they noted that that collaboration on preserving the delta is already ongoing between the two countries in scientific, academic, and nongovernmental organizations; and they noted that "some studies conducted by these groups" have defined the ecological issues in both countries regarding the delta. Most important, they noted that "entities in their respective countries may seek water and seek to ensure its use for ecological purposes in . . . the Colorado River delta" (2).

That last sentence makes the document remarkable—water for the delta is now formally and explicitly part of the law of the river. There is nothing binding here—only the statement of a long-term goal, or possibility. But there it is.

No water is actually allocated to the delta. Instead, the conceptual minute establishes a process for determining exactly how much water would be needed and how it might be provided. The minute calls for systematic and cooperative studies that "include possible approaches to ensure use of water for ecological purposes in this reach [of the delta] and formulation of recommendations for cooperative projects, based on the principle of an equitable distribution of resources." The minute also calls for a "binational technical task force" to "examine the effect of flows on the existing riparian and estuarine ecology of the Colorado River from its limitrophe section to its delta, with a focus on defining the habitat needs of fish, marine and wildlife species of concern in each country" (4).

It is called Minute 306.

This development is a huge step forward. Biologists such as Ed Glenn and Karl Flessa, among many others, are developing a comprehensive research plan that will expedite matters enormously on the scientific front. In September 2001, the stakeholders in both countries met in Mexicali again to learn about the laws governing the river and the studies done on the delta and Gulf of California. At this meeting, the Mexican delegation adopted a document that identifies sixteen specific courses of action that Mexico can take to move the process forward toward a formal minute that guarantees water for ecological purposes in Mexico. The document notes that the treaties entered into by both the United States and Mexico "do not include the environment, and specifically the Colorado Delta and the Upper Gulf, as a user of the Colorado River basin, and therefore this circumstance should be remedied in the future in a shared forum with joint responsibility" ("Courses of Action" 2001, 1).

In addition to this pledge of joint responsibility, the Mexicans asserted that "both governments promise to provide volumes of water to protect and restore the ecosystems of the Delta," and that "both countries adopt the principle that they will not take unilateral actions that limit future actions for the restoration of the ecosystems of the Delta" (2).

The document adopted by the Mexican delegation is a sign of the growing commitment within Mexico to ensure that any water reaching the delta for ecological purposes will be used for fish and rails, not farm sprinklers and urban toilets in Sonora and Baja California.

Since these meetings, the presidential election in the United States brought in a new secretary of the interior, Gale Norton, who at this writing appears decidedly less sympathetic to the environment. Nevertheless, environmentalists are hoping that President Bush's close relations with Mexican President Vicente Fox will help move the process forward. Meanwhile, the seven basin states have closed ranks, skeptical and increasingly worried. Battle lines are being drawn as armies of lawyers marshal their arguments to stymie any claims against their precious water.

In any event, the conceptual Minute 306 adds a new environmental dimension to the body of law governing the river, and brings us one giant step closer to the larger dream of the delta's restoration.

It is no surprise, as you study the new water wars in the West, to discover that the origins of the English word "rivals" comes from the Latin word *rivalis*, which means persons who live on the opposite banks of a river used for irrigation. Rivals are fundamentally, etymologically, competitors for water. The states may like to compete among themselves for Colorado River water. But historically the chief rival has really been Mexico. The conceptual Minute 306 signals a significant shift of diplomatic momentum on behalf of the delta. But achieving an actual formal minute that involves international water transfers for the Mexican delta will be very difficult.

Such transfers of water make bureaucrats on both sides of the border nervous. The states have declared their opposition in a number of forums. Perhaps the most succinct summary of water-user opposition in the United States was expressed at a conference on the law of the river. Tom Levy, representing the Coachella Valley Water District in California, responded to concerns about the delta by denying that it is even an issue. "We solved the delta problem in 1944," he claimed, referring to the international treaty. "This is not a U.S. problem. It is a Mexican problem."

Such language perfectly expresses the closed mindset against the delta. Getting the seven states in the Colorado watershed to let go of water for the environment in the United States has been grueling. But to give water to Mexico for rails and fish, wetlands and cottonwoods—well, that's nearly unimaginable, tantamount to a sacrilege of the West's most precious resource. If these states can agree on nothing else, they can agree on this: no more water for Mexico.

Whether it is the Multi-Species Conservation Program (MSCP), the lawsuit under the Endangered Species Act, or a new conceptual minute to the 1944 treaty, the states generally want no part of it.

Apart from appeals to states' rights on the river—though they have been heavily subsidized by the federal government and the Bureau of Reclamation—the states make two arguments. Both arguments try to lay the burden of environmental restoration in the Mexican delta on Mexico.

The first argument concerns Mexican sovereignty and says that the United States has no power to tell Mexico what to do with water within its borders. If the United States were to dedicate water for ecological purposes in Mexico, the argument goes, we can't guarantee that Mexicans would not simply use the water for, say, farming. The pressures for water are at least as great in Mexico as they are in the western United States. Mexico has shown no desire to dedicate its own water to environmental restoration. Why should anyone in the United States assume that they would do so with new water supplies?

In testimony to a U.S. House committee meeting on Colorado River management issues for the twenty-first century, Kent Holsinger, assistant director for the Colorado Department of Water Resources, submitted a statement entitled "A 7-State Perspective." He maintained:

*Yellow-headed blackbirds croak from reeds throughout the wetlands of the lower delta.*

> The Congress should be aware that all water resource use on the
> Colorado River within the United States has been consistent with
> the 1944 Mexican Treaty and other aspects of the 'Law of the River.'
> The State of Colorado stridently objects to any suggestion that water
> for the restoration of Mexico come from the Colorado River in the
> United States. We urge the Congress and the Bush Administration
> to ensure that any discussions regarding the Colorado River Delta be
> done in full and complete consultation with the seven Colorado River
> Basin States.

The statement goes on to claim that the delta is Mexico's problem:

> . . . The previous administration initiated discussions with Mexico and
> environmental organizations to address perceived water needs of the
> Colorado River Delta in Mexico. These efforts resulted in a new conceptual
> Minute (306) to the International Treaty with Mexico. . . . The State
> of Colorado has serious concerns . . . and believes Mexico should be
> responsible for water issues within its own borders. If environmental
> issues need to be addressed, we encourage Mexico to pursue flow
> management and structural alternatives within their borders.
> (Holsinger 2001, 2)

Apart from the strident hostility of the tone, Holsinger's point is simply that the delta in Mexico is not the United States' problem. It is certainly not the seven states' problem. It is Mexico's problem and Mexico needs to solve its own problems with its own water.

The states form a powerful phalanx of forces when allied with water districts, irrigation districts, and big cities—the water users of the United States. In an amicus brief on behalf of state water boards in Arizona and California, as well as various irrigation districts and the cities of San Diego and Los Angeles, lawyers make the case on Mexican sovereignty and culpability even more explicitly. They filed the brief in opposition to the lawsuit brought by the Defenders of Wildlife and many others under the Endangered Species Act. Virginia S. Albrecht and two other lawyers write in their brief to the court:

> The choices about the proper allocation of water in Mexico are exclusively for the sovereign nation of Mexico to make. . . . Mexico diverts and uses virtually all LCR [Lower Colorado River] water that enters its borders for agricultural and municipal use. In addition, these waters are subject to other uses and impacts, including recreational and commercial fishing, pollution from agricultural run-off and other sources, and natural forces including evaporation and absorption. . . . Federal Defendants [Bureau of Reclamation and Department of the Interior, etc.] reasonably and correctly concluded that effects on species in Mexico associated with the waters of the LCR are attributable to Mexico's complete sovereign control and use of that water once it enters Mexico and actions by persons in Mexico, not Reclamation's actions. (Albrecht et al. 2001, 1)

Again, if there is a problem with the delta, the argument goes, it is Mexico's fault. The real reason these water users oppose even a drop for Mexico's delta has little to do with the law. It has to do with economics and growth. Take Colorado. It has not used its full allocation of water yet under the law of the river. A part of the law, the Upper Basin Compact of 1948, divides the Upper Basin's 7.5 million acre feet not by strict volumes, but according to percentage of flows contributed to the river. Colorado is the big winner with 51.75 percent. Utah gets 23 percent, Wyoming, 14 percent, and New Mexico, 11.25 percent.

The water not yet used by these states is being guarded for future growth. Colorado expects to grow significantly in the next twenty years. As Holsinger says, "The Colorado River Basin offers the only significant source of new developable water for the State with approximately two million acre feet available under the Colorado's compact apportionment" (Holsinger 2001, 1).

Some say that the cities are the ones to watch as the debate goes forward. The cities will get their way. The old agriculture interests will be pitted against the environment. Yet when it comes to Mexico, a familiar coalition of states, cities, and irrigation districts—the insatiable users—want their own way. They would have us wash our hands of all responsibility for the death of the delta.

The future, then, would be the delta's recent past, tragically repeating itself. The last great drought in the delta came when the reservoirs of the dams filled during the twentieth century, and the delta may be in store for a second dry spell.

Whatever the ulterior and even selfish motives of the states, the question of Mexican sovereignty is central to the delta problem. It was certainly on our minds as Bill Snape and I stopped into the Mexicali offices of Francisco Bernal, an engineer who represents Mexicali water interests on the Comisión Internacional de Limites y Agua (CILA), the Mexican counterpart of the IBWC. It's exactly the question we want to talk to Francisco about.

Francisco is polite, but he does not look like he really wants to talk to us. He is nervous, fidgety. A technocrat caught in the middle of heavy-duty politics, he is not in his element. He asks us immediately why we want to talk to him. We try to be reassuring, chatty. But he sits in his air-conditioned office, lined with huge maps of the border, and tries to change the subject.

What he wants to talk about is not policy. He wants to talk about technical issues, decidedly a comfort zone for bureaucrats. "I'm not the policy representative," he says. "I'm field coordinator here in the Valle de Mexicali." That's the southern part of the fertile delta valley that includes the Imperial Valley in the United States, just a stone's throw away across the brick wall and cyclone fence border from where we sit.

"We are in the process of developing a working agenda on the water issues of the delta," he says. "But that's the big problem now, reconciling the ecological interests with those of farmers and cities."

Francisco outlines the looming water issues Mexico faces, apart from the ecological matters. The two major Mexican cities in the region, San Luis Río Colorado in Sonora and Mexicali in Baja California, have boomed with the increase of international investment. This growth is partly NAFTA related. More though, it derives from the fact that Mexico has courted Asian investors, notably Korea and Japan. In 1900, Mexicali was just a one-cow outpost. It quite literally does not exist in historical population records of Baja California. In 1970, Mexicali's population was 277,306. By 1998, that number had tripled, to 745,027 people.

"Both Mexicali and San Luis Río Colorado are growing at 3 percent to 4 percent per year," Francisco tells us. "Tijuana is growing at 6 percent per year." For Mexicali and San Luis Río Colorado, the Colorado River and groundwater are the only sources of water. Even Tijuana gets water pumped across the peninsula from the Colorado River.

The *maquiladoras* have been the heart of a neoliberal, progressive policy of economic growth along the border, following the formula of cheap labor and foreign investment. But these are desert cities. "Water resources are not growing," says Francisco.

As for agriculture, Mexico's 1.7 million acre feet allocation of the Colorado River irrigates almost 500,000 acres in the Mexicali Valley. That's just slightly under the acreage irrigated in the United States' Imperial Valley. But Mexico's water resources are more limited. Where the Imperial Valley is able to plant two or more crops per year, Mexico plants only one. And the Imperial Valley is able to build an elaborate system of subsurface drainage tiles—an engineering marvel—so that freshwater can be pumped through the soil to flush out accumulating salts, keeping the soil productive. Mexico has neither the capital nor the water for anything equivalent.

Francisco tells us that Mexico is, in fact, using almost 2.5 million acre feet in the delta, all of it allocated to farms and cities. The difference between this and the Mexican allocation of Colorado River water (1.5 maf, plus 200,000 af, if available) is made up by pumping ground water. Francisco swears that regulations are keeping Mexico from depleting the aquifer. But others I have talked to say that the Mexican government does not release the statistics on this and that many are sure the aquifer is being depleted.

From Francisco's perspective, the most pressing water issues for this part of Mexico do not include the ecological issues in the delta. For most Mexicans the biggest is the lining of the All-American Canal. This canal closely follows the international border, delivering water from the river to the Imperial Valley in California. It got the rah-rah name because the canal replaced the original diversion route that went through Mexico and then back north into the United States, a scheme that left American developers feeling too vulnerable. When technology, combined with federal funding, made a new canal possible through these dunes through U.S. lands, it was built and christened, patriotically, the "All-American Canal."

The All-American Canal is a huge ditch of water that flows straight and true through the desert to the rich farmlands of the Imperial Valley. It is not lined with concerete, it courses over porous sands, and it seeps into the aquifer. On the other side of the border, Mexican farmers have drilled wells, *pozos*, to tap this water source in Las Dunas, the sand dunes near Algodones. Farmers have 121 wells in Las Dunas, affecting about 33,000 acres of farmland.

Here's the rub: The water conservation measures for the California 4.4 Plan call for lining this canal, saving some 67,000 acre feet for southern California cities—and putting these Mexican farmers in jeopardy, costing them an estimated $80 million per year.

Francisco tells us that throughout the Mexicali and San Luis Valleys there are over seven hundred ground wells, pumping almost half a million acre feet from these wells.

Mexico has its own card to play here, though. Not every river in the lower Colorado River watershed flows south into Mexico. Some flow north from Mexico to the United States.

One of them, the New River, flows through Mexicali and empties in the United States' Salton Sea about forty miles away. It is a notorious river, often called the most polluted river in North America. Mexicali dumps its untreated and inadequately treated sewage into the New River and lets it drain north in a foul and frightening flow.

In 1980, Mexicali and Calexico, the U.S. city just across the border, agreed to clean up the New River. "We've spent almost $7 million to improve the system," Francisco boasts. A few more million and they will have a Mexicali sewage treatment system ready to come on line. And the clean water it will produce? Where will it go? The United States? The Mexican farmers? Is it a bargaining chip, as a rumor has it, to keep the United States from lining the All-American Canal?

Francisco is coy and noncommittal. He says laconically, in response to my question about what will happen with this water, "We have no agreement to send water to the United States."

It is not clear that Mexico intends to use that water for, say, cleaning up Río Hardy, as people like Monica Gonzalez of the Cucapá and Javier Mosqueda of a water user's association have hoped. So will Mexico use new water from the United States for ecological purposes? The politics of this will be, as José Campoy admitted to me in a conversation, "strong." It is quite clear that Mexicans do not want Americans telling them what to do in their country. They have seen enough of that in the past. It is also clear that, while every water user in Mexico would love more water for their own gains, they realize that more water for the environment is a gift they cannot turn down.

Nor is the idea to dictate to Mexico what it does with its water. The effort to restore the delta is not an isolated conservation phenomenon. It is part of a long process of binational efforts to address ecological issues along the border—issues that are insoluble without binational cooperation. The following is a list of the most significant parts of this developing multinational process, part of gathering historical momentum on behalf of the environment:

1936: Convention for the Protection of Migratory Birds and Game
Mammals (a hemispheric agreement)

1941: Convention on the Nature Protection and Wildlife Preservation
in the Western Hemisphere ("Western Hemisphere Convention")

1971: Convention on Wetlands of International Importance Especially
as Waterfowl Habitat ("Ramsar Convention")

1983: Agreement between the United States of America and the United
Mexican States on Cooperation for the Protection and Improvement
of the Environment in the Border Area ("La Paz Agreement")

1986: North American Waterfowl Management Plan

1992: Integrated Environmental Plan for the United States–Mexico
border area

1993: Creation of the North American Commission on Environmental
Cooperation

1994: Canada, Mexico, and the United States Trilateral Committee
for Wildlife and Ecosystem Conservation and Management
("Trilateral Committee")

1996: U.S. Border XXI-Frontera XXI (establishes binational working groups
and provides funds to address environmental issues in the border region)

1997: The Letter of Intent between the Department of the Interior of the
United States and the Secretaría del Medio Ambiente, Recursos
Naturales, y Pesca (SEMARNAP, now SEMARNAT) of the United
Mexican States for Joint Work in Natural Protected Areas on the
United States–Mexico border

2000: Secretary of the Interior Bruce Babbitt and Julia Carrabias, secretary
of SEMARNAT, sign a Joint Declaration between the two countries,
pledging binational cooperation for saving the delta of the Colorado
River. Finally, the conceptual Minute 306 of December 2000 also
testifies to the binational working relationship between the two
countries.

One should also add that in 1993 Mexico took the significant and even
internally controversial move of creating the huge Reserva de la Biósfera Alto
Golfo y de California Delta del Río Colorado.

Mexico in general does not have the same history of environmental awareness found in the United States. Such awareness is still in its nascent stages. In
fact, the delta as an environmental issue seems to be helping Mexico move
toward a new national sensibility in favor of the environment. The delta has
been the subject of several major news features, for example, in Mexico City
newspapers. And despite internal political pressures, Mexico has shown every
indication of working with the United States in finding a binational way of

saving the delta. That will be the job of a formal minute to the 1944 water treaty: allocating water to the delta's restoration.

There is one crucial difference, however, between these two water nations: the amounts of water available to the respective countries. This makes all the difference in their relative prosperity, because water is money.

The issues on both sides of the border mirror each other. Most people recognize the importance of finally addressing the long-neglected

ecological issues throughout the watershed in a positive way. Yet the bureaucrats and users worry about practical matters of supply. These are two water nations in a desert and they need every drop they can get. At the margins, the environment is left to battle with agriculture, while the cities keep growing, chugging down more and more water. One other apparent parallel between the two countries is that so far, the federal government of each country seems to be working on behalf of the environment and trying to bring the users and states into line.

The time is soon coming, however, when the Mexican government will have to demonstrate its commitment to the delta in clear language. It will have to declare its intentions in unambiguous terms in order to make a transboundary water minute possible.

Francisco Bernal does not raise one's confidence in the commitment of Mexicans inside the government to save the delta. It is probably more accurate to see him, though, as a Mexican equivalent to one of the water managers in the United States. Uncomfortable as he was in saying anything that might compromise himself, he acknowledged the new voices raised on behalf of the delta. As we leave his office, Francisco says, "Lately water amounts have been high. People are happy. Species are happy. I like the ducks and the animals too."

*Osvel Hinojosa, graduate student at the University of Arizona, conducted the rail census throughout the Mexican delta. In addition to discovering many overlooked natural areas in the delta, he works to find a solution to the problem of water in the delta.*

✿

I believe animals help humans. They teach us. They show us what we need to learn. A few days after visiting with Francisco Bernal, I drive back into the delta and I am with animals, who teach me a new lesson about the delta.

A mud levee separates the vast mudflats and wild portions of the delta from the settled areas. It is a vast earthen boundary over ten feet high that provides

The burrowing owl is a common resident throughout the delta, especially on the levees that separate the farmlands of the delta from the lower, "wild" regions to the south.

flood control—from the river or from high tides—on the delta. You can drive up on top of it. In fact, the Mexican military regularly patrols it. Mexicans call it *el bordo*, a big raised furrow. It is evening and I'm doing one of my favorite things to do in the delta: looking for burrowing owls.

Burrowing owls, *Athene cunicularia*, are barely ten inches tall. They are abundant and one of the comical creatures of the delta. They stand on spindly legs very upright, have buffy backs and light chests. Their upright, two-legged posture makes them look amazingly human. Their large eyes and rounded faces contribute to the effect. Their eyes are bright yellow. They stare back at you with a withering and distrustful intensity.

In delta Spanish, the birds are called *tecolote*, generic for owl. The word gets applied to different species of owls in different parts of Mexico. Juan Butrón taught me the local name for these creatures: *agachones*. *Agachar* means "to squat."

The burrowing owl is the squatter, and it often makes me laugh as it squats and squints, looking back at me. That is one of the characteristics about burrowing owls I love the most. They squat or bob as they look at you with those cannon-like eyes. They are taking your measure.

This evening, I am out just as the sun is fading. The brown earth is taking on a biege, grayish quality as the shadows creep up. I am trying to use the car as a blind, driving along the levee, to get close to a family of burrowing owls and photograph them. I find two small owls standing sentry over their burrow, as these creatures typically do. They are preparing to head out for a night of hunting. They let me approach, as they squat nervously and then fly off with an alarmed screech. They have to do this a lot because they have chosen to live on what is essentially a road, a very rough road, ringing the delta. Cars often drive right past the eroded holes in the levee out of which they make their burrows.

A ray of pinkish light highlights one of the owls as it lands a short distance away, on a clump of mud farther up the levee.

The bird eyes me as I get out of the car and inch toward it. I move slowly. This *agachon* bobs and flies. But finally it lets me come very near. It is in soft yellowish light and it lets me photograph happily. The owl sits on its gangly legs, bobbing up and down at me.

This is when I realize that it has something to teach me tonight about seeing the delta and managing the delta. Burrowing owls bob because they do not otherwise have a good way to measure distances. By bobbing, they are able to take a double reading on an object, say me. They create an angled vision. The squatting enables them to judge depth and distance. They get two perspectives in their vision. They see two times.

It is called the parallax method of seeing. Astronomers use it, for example, in order to measure the distance to a star. They take a measurement of the star from two angles and then can figure out its distance. It is a method of determining the viewer's relationship to an object of sight.

The burrowing owls squat and bob and give me a lesson in the value of seeing twice. These creatures live on the boundary between the wild and the cultivated. They do well in both. We humans could learn a lesson in double vision from burrowing owls. We need our own double vision for the delta, to see it both as wild and as cultivated. Both are important. Wild and cultivated, Mexican and American: this is the parallax method of seeing applied to the delta. Ironically, to preserve the wild, we have to cultivate it.

Before leaving for the night, I bob back to the almost comical burrowing owl. It is a gesture of thankfulness for a lesson in seeing twice.

TWELVE

# The End of the River

ENDEMIC SPECIES: Thick-billed savannah sparrow and Palmer's sea grass
PLACE: Colorado River mouth

THE *PANGA* SKIDS wildly down the slickened mud bank and slams backward into the water. Bernabe Rico Olague jerks the attached rope and jumps quickly aboard. The river is rising fast and the boat is drifting into the swift currents as he tries to start the motor. José Campoy, the director of the biosphere reserve, steps gingerly down the gooey mud, careful not to slide into the river. He waves for me to follow. Together at the very edge of the river, we watch the rising waters and wait for Bernabe—nicknamed Berna—to get the damned engine going.

Only half an hour ago, it did not look like we would get onto the river today for our *recorrido*, our excursion, to Isla Montague. We had already tried once, days ago, and the wind was too violent, the waves too dangerous. Today, we arrived to find the tide going out and the river low in its channel, twenty-five feet down the crumbling mud bank. Even as José shook his head, the tide began to change. It began to rise hard. I knew the delta tides were famous for being among the highest in the world. As the water came surging up the bank, I glimpsed something of the force of the ancient bore, *el burro*, at the mouth of the river. When the incoming tide was high and the river running strong in the spring floods, the two opposing currents would meet in a clash of water at the mouth of the river. They created a wall of water that moved backward up the river.

*El burro* was once strong enough to capsize the thirty-six-ton steamer, the *Topolobampo*, killing eighty-six peasants from Guaymas. Only twenty-one bodies

◄ Pasto gentil was harvested by the Cucapá Natives. The wonderful grass on Isla Montague was adapted to salt water and a spring flood of fresh water from the river.

▲ Shorebirds like this snowy plover nest in the mudflats of the lower delta. It is on behalf of creatures like this that conservationists want to protect and restore the delta.

233

were found. Only thirty-nine people survived. Days after the wreck, survivors were still being dragged off the mudflats around here, wild-eyed and naked, half mad with thirst and the searing sun, savaged by predatory swarms of insects. (Waters [1902] 1974).

*El burro* was one of the river casualties when dams tamed the Colorado.

What we are watching this morning is not really a rising river, but a rising tide. Though the mud-thick river at our feet looks large, it is swollen not with

freshwater but a with a high tide surging in from the Gulf at an astonishing speed, in the sludgy brown chop of waves.

Water in the American Southwest is supposed to be able to defy the laws of nature: it flows uphill toward money. Here in Mexico we found a literal example, as we watched the laws of money work on the laws of nature. The water in the delta has been long-gone to interests upriver. The Colorado near its mouth is a river running in reverse.

*When the Colorado River finally reaches the Gulf of California, it breaks in two and flows around the low muddy island of Isla Montague.*

Now the Mexican delta has managed to insert itself into the smack-down world of water politics in the United States and Mexico. The likelihood of success for getting more river water that might someday actually meet the surge of tide in the Mexican delta depends entirely on how effectively the delta's champions make their case in the United States, amid a growing list of very demanding water interests.

This morning I am beside the river with José Campoy, the man who is perhaps most deeply involved in these difficult political crosscurrents of real water in Mexico and legal crosscurrents in the United States. As director of the biosphere reserve since 1996, José is a biologist turned delta champion. When he began, it must have seemed an almost impossible job. Hardly anyone gave a damn about the delta. Except for a few vagrant rumors among a limited *cognoscenti*, the delta had long since been written off, sadly forgotten in the United States by the people who enjoy and depend upon what was once its water.

José's wife, Martha Román, works for the Instituto del Medio Ambiente y Desarrollo Sustentable del Estado de Sonara (IMADES), a conservation agency out of San Luis Río Colorado, where José and Martha live with their family. Their two small children often visit the biosphere reserve's office and Martha is also active in delta restoration.

José is as affable a man as you could wish for. He looks almost boyish, and is eagerly accommodating, within reason. They are qualities that serve him well in his sensitive political position, because to direct the biosphere reserve means he lives in the center of controversy. Trained at the Universidad Autónoma de Baja California in Ensenada as a fish biologist, José began working in the delta studying *pupos*—looking for relict populations of the fish in isolated *pozos* or springs. Since becoming director, he travels all over trying to convince both Mexicans and Americans that the delta is worth saving. And he tries to find ways of getting the water that will make that salvage operation possible.

José still manages to get into the field, though more rarely now. Today we are on our way down the river to Isla Montague, the almost fantasmic island that sits like a muddy plug in the very mouth of the river. The trip gives us a chance for us to talk delta politics.

"So little water left in the river," José says, as we stand on the mud, water rising. "We know how difficult it is going to be to get the United States to give up more water. We know all the threats in the United States to helping the delta. For a long time people felt the delta did not deserve recognition. People in the United States say, 'We don't know the delta.' But when they discover it, they want to help. They see it's damaged and needs restoration. I always say, that's the challenge. That's the hope."

So we are going on our own *recorrido*, or tour, on a damaged estuary—to the end of the degraded river.

Berna finally gets the motor to cough and start and we jump dextrously in, and then brace ourselves. Berna is a ranger for the biosphere reserve. He works mostly at night, patrolling the waters and looking for people fishing and shrimping in violation of reserve laws. His huge head of black hair and wraparound sunglasses set off a handsome face. He loves to tell wild stories of his adventures in the delta, casting himself as the wild man on the river. In his stories, he jokingly calls himself *pinche Berna,* loosely translated as "that damned Berna."

He guns the boat out into the currents, which are coming hard from various directions. The boat jolts across the chop of waves. Berna charges into a full-throttle, breakfast-banging, white-knuckle ride twenty miles down the Colorado River to its mouth.

Near the mouth, the river seems to widen. In the low flats of mud and water, unrelieved by plants or geological features, it is very hard to tell exactly where the river ends and the sea begins. Isla Montague makes the task more difficult. Here the river is reduced to primal essentials—water, mud, and sun.

José points ahead and yells. "Isla Montague!"

We leave the river channel and enter the spot where sea and river wrap around the island. To the east lies the Sonora shore; to the west, Baja California. Crossing the widening water, we skirt the north end of the *isla.* Berna scouts out a small drainage crease in the mud of the island and nudges the boat into a small but protected slough.

While Berna works on the motor, which is acting up again, José and I take a walk across the flat island, which barely rises above the water where the river meets the Gulf. In floods, of course, this whole island would be inundated—submerged.

I look around me at the river and the sea and the muddy Isla Montague. I take advantage of the moment to suck in the scene. This is not a place that appeals powerfully to the visual sense. Its overwhelming feature is flatness—water and sea below, flat sky above. The entire landscape, even the island itself, seems profoundly insubstantial, really just water and mud and light falling from above, bouncing from below. At the island's southern end, there is a small attached island, Isla Gore. José calls it a *fantasma.*

Such is the mysterious, phantom nature of this place. It is a dreamy, fantastical landscape of waves and muddied boundaries, of light waves and water waves. Even the island itself is slippery, more water than earth. It's the liquefaction of the landscape. Suddenly the ancient Greek philosopher Thales seems right. Water is the only true thing, he taught, the one reality behind all other appearances. That belief is more than metaphysics here on the river. It is a political fact that everyone knows in their bones.

For hundreds of years, the early Spanish explorers to this area had trouble understanding the geography, in knowing what they were actually seeing. When

*José Campoy (left) is the Director of the Biosphere Reserve of the Upper Gulf of California and the Delta of the Colorado River. Bernabe Rico Olague is a warden for the biosphere reserve.*

they tried to map this place, they could not find its dimensions or measure its spaces. So the area became part of a mythic, literary European landscape. Hernán Cortés himself traveled to the Baja peninsula in the 1530s, thinking that he was traveling to an island in the great ocean—an island inhabited only by Amazon women. His own captain, Francisco de Ulloa, sailed up to the delta in 1539 through the "great gut of water," which he named *el Mar Bermejo*—the Vermilion Sea—and found the river right where José and I are standing.

Cortés' rival for power, the first viceroy of Mexico Antonio de Mendoza, sent his own representative, Hernando de Alarcón, up the Gulf in 1540, hoping that Alarcón would meet with Francisco Vásques de Coronado. They missed each other, but Alarcón actually sailed up the river, perhaps as far as the Gila.

Shortly after de Ulloa and Alarcón visited, the delta sank back under a sea of dreams. The Baja peninsula became linked with mythic places. Geographers considered it an island. Drawing on stories from the enormously popular chivalric Spanish romances of *Amadis de Gaul*, they called the island "California," home of the Amazon warrior queen Califía.

For two hundred years, the delta was not the delta. It was part of a fantasy island of California. Only the explorations of Father Kino proved that the delta was not an island, when in 1702 he entered the delta from Sonora, descended nearly to the mouth, and crossed the river in a basket accompanied by Indians. "California no es isla," he declared. California is not an island. His exploration set the stage for the Spanish colonization of California.

Yet as we walk the island, it is easy to see how this land could be so dimly perceived. It offers little upon which to anchor oneself. Yet it also seems that modern Americans and Mexicans have not done much better than our predecessors in understanding the delta for what it is. Though we imagine ourselves as hardheaded, practical, realistic people with clear-eyed vision in an empirical age, we have imposed our own fantasies here on the delta.

For us, though, it has been a fantasy built out of dreams of conquest. Our privileged texts are not great chivalric poems of the Middle Ages. But they are texts no less equally revered by us—they are the texts that form the law of the river. We have used the language of the law to occlude our own vision, which in turn serves our own cultural agenda. Projected onto the actual river, the river of law has completely eclipsed it and its delta.

But now we are being given a chance for a kind of second vision in the delta. A new geography is taking shape in the minds of people on both sides of the border, more fully grounded in actual lives lived in this place and the realities of this landscape.

José and I leave the boat and take off on a walk across this flat island; it's a horizontal world. It's hard to get a real grasp on the sense of space. Because where we are seems to have no boundaries, it feels empty, a bit disorienting and strange, undifferentiated in every direction.

The *isla* is covered in a sparse and prickly grass, through which José and I walk in our shorts. It stabs at our bare legs like needles. A small sparrow flushes from the grass at our feet. "Savannah sparrow," José says right away. "But it's a subspecies that lives only in the delta—an endemic found only here. It's called the thick-billed savannah sparrow. It nests on the island."

So do many of the seabirds flying overhead. I note the Caspian terns, with wild dark crests and blood-red bills. Then a rarer tern, a royal tern, cruises past. I watch for black skimmers, one of the most wonderful and bizarre of the seabirds. It's a like a tern with a huge red bill. It glides along the water with its lower bill open, skimming the surface and slicing the water to gather small fish—hence its name. Isla Montague is a favorite nesting spot for black skimmers, but I see none. Then a rare yellow-footed gull, another great specialty of the delta and Upper Gulf, flaps by overhead.

José picks a stalk of the grass and shows me the white crystals clinging to its stem. "Crystals of salt," he explains. "It excretes salt. It's *pasto gentil*, Palmer's sea grass. This is the plant the Cucapá harvested. It only grows here in the mudflats of the delta. It only needs one big flow of freshwater from the river every year. Then it can grow in salt water."

The adaptations of life to this environment seem almost miraculous. That the Cucapá could have adapted to this place, could have humanized it, seems no less miraculous and wonderful. The place suddenly begins to take on dimension.

We go back to the boat. Berna is still tugging at the motor.

José's hope is most profoundly aimed at obtaining a new formal minute to the 1944 water treaty so that water can be dedicated to the delta for ecological purposes. He might not have fully realized, when he took this job as director of the biosphere reserve, that he would be so immersed not only in Mexican water politics, but U.S. water politics as well. He knows the task is to convince people in the United States—states and various users who are resistant to providing water for the delta—to learn about the area, to reach out beyond the boundary, to be more inclusive in their vision.

He would love to take part in discussions of water policy in the United States, for example the Multi-Species Conservation Program. But he is excluded. The best method for understanding, he believes, is to get people down into the delta to experience and see what is really here.

"There are big threats to the delta now," José says. He is terminally optimistic. Perhaps it is a requirement of his job, but it serves him well. "Very big threats. All the issues with water above the United States border. But when I began on the reserve, the Bureau of Reclamation said they had no interest in the delta. Now it's very different, in a positive sense. On both sides, the United States and Mexico. People are coming to the delta and discovering what's here. It makes me have more hope than ever before."

José would love people to see the delta as connected to the rest of the Colorado River watershed. To do that, though, he knows that he must understand the U.S. users' resistance to the delta. The entrenched attitudes go very deep. The heart of the legal opposition to sharing water derives from questions of Mexico's national sovereignty and cooperation across borders. Plus, José knows he must convince people in Mexico to make the delta a conservation priority. Creating a formal minute is part of creating this new delta geography.

Berna gets the motor going again. We all climb in and head back upriver. As we do, I hold on tight. Like the politics of the river itself, the ride is a bumpy one over choppy waters and hard-to-read currents.

※

From the beginning of my interest in the delta, one question has haunted me, a question that goes right to the heart of creating a new vision and a new geography in the delta.

I asked the question of everyone I thought might have an answer, Mexicans and Americans. No one had a good answer. Few would even hazard an answer. Most wrote my question off as unanswerable. Yet it is central in the debate about providing legal guarantees for transboundary deliveries of water for ecological purposes in the delta. For many people the 1944 water treaty is viewed as settling a century of bitter battles—literal and diplomatic—over the boundary between the United States and Mexico. Yet the status of this boundary, however legally codified, remains more fluid than many would like to admit.

It seems foolish that Mexico signed the 1944 water treaty and left themselves with only a 10 percent cut of the river allocations while the United States gets 90 percent. Mexico is even more strapped for water for agricultural and urban needs—much less ecological restoration—than the United States.

So the question is: Why did Mexico sign a treaty that left it with so little of the river?

There are ethical reasons that the United States should help restore the delta. There are also practical reasons, since Mexico has so much less water to work with for ecological restoration. Water is tight on both sides of the border. Yet Mexicans repeatedly point to the water waste and water luxury in the United States: green lawns and city fountains and golf courses. The inequity extends to farming as well. As one farm historian in the delta, Oscar Sanchez of the Universidad Autónoma de Baja California, Mexicali, told me, in the Imperial Valley farmers have enough water for two crops per year. In the Mexicali Valley, farmers have enough water for only one crop per year. Beyond ethics and practicalities, the answer to this treaty question might provide an historical reason that the United States should help.

At first I wondered if Mexico did not negotiate well. I now believe, however, that the answer at least in part is that Mexico has historically always fought from an inferior and disadvantaged position along the river. Our southern neighbor has consistently been at the mercy of the United States' superior technology and superior resources.

Mexico was never in a position to exploit the resources of the Colorado River on its own. It has only one dam on its side of the border, the Morelos Dam near Yuma, built as a stipulation of the 1944 water treaty to enable delivery of treaty waters. You will not find other places in the deep sedimentary soils of the delta—a land without bedrock—where you could even anchor and build a dam. Additionally, for Mexicans the delta is tantamount to a very hot Siberia, a long way from the centers of power in Mexico City. For the United States, the Colorado River is in a forbidding desert, but within easy striking distance of Los Angeles.

The distribution of Colorado River water outlined in the 1944 water treaty clearly represents an agreement between two sovereign nations. The treaty is also an historical document that defines relations at a particular point in time. And in a sense, it is a legal codification of the power differential that existed between the two countries at the time of negotiation and signing. Mexico's delta grew up dependent on the development on the U.S. side of the border, becoming essentially a hydrological community of the United States. I developed this understanding from reading Mexican histories of the water treaty with the United States, particularly Ernest E. Coyro's massive two-volume work, which I will explain below. It is also an extension of Donald Worster's provocative and largely accepted thesis, *Rivers of Empire: Water, Aridity, and the Growth of the American West.* Worster argues that building the dams in the West's deserts was a project in empire building. I would extend this thesis to the U.S. relation with Mexico in the Colorado's delta, which is not examined in Worster's book.

Worster's "hydraulic hypothesis" explains the vital relationships between water, culture, and power. Worster argues:

> The American West can best be described as a modern hydraulic
> society, which is to say, a social order based on the intensive, large-scale
> manipulation of water and its products in an arid setting. That order is not
> at all what Thoreau had in mind for the region. What he desired was a
> society of free association, of self-defining and self-managing individuals
> and communities, more or less equal to one another in power and authority.
> The hydraulic society of the West, in contrast, is increasingly a coercive,
> monolithic, and hierarchical system, ruled by a power elite based on the
> ownership of capital and expertise. (Worster 1985, 7)

Power was deployed through huge water projects throughout the West, a history that has been well documented as a travesty of the original plans for small-farm homesteading. These water projects served different large state and private capital interests.

They also derived from an "imperial" mentality, hence the book's title, *Rivers of Empire.* In Worster's opinion, it was no accident, that the first major water diversion of the Colorado River went into southern California's aptly rechristened Imperial Valley in 1901. What was being acted out was more than a desire for water—more than engineering wet dreams, if you will. To Worster, the dams and diversions in the desert were "compelled by an unrelenting American cultural imperative" (Worster 1985, 188). Like the modern space program, this cultural imperative was so great it could not be shaken by failure or risk. It derived from a deeply seated mentality, a psychological drive. To explain the imperative, Worster quotes John Widtsoe, who served with Elwood Mead, the chief U.S. negotiator on the 1944 water treaty. Widtsoe says emphatically:

> The destiny of man is to possess the whole earth; and the destiny of the
> earth is to be subject to man. There can be no full conquest of the earth,
> and no real satisfaction to humanity, if large portions of the earth, remain
> beyond his control. Only as all parts of the earth are developed . . . and
> brought under human control, can man be said to possess the earth. The

United States of America . . . might accommodate its present population within its humid region, but it would not then be the great nation that it now is. By the vision of its statesmen and by its marvelous power of accomplishment it has made use of the country west of the hundredth meridian, which lies under low rainfall. The nation is now one country. . . . And all the world will be helped by the conquest. It is an imperial problem which as it is solved will satisfy a world-wide need. (Worster 1985, 188)

Worster concludes that "total power, total possession" was the program (188). The great dams and the irrigation projects were instruments of both wealth and empire through the conquest of nature. But he does not address how this program affected U.S. relations with Mexico, with whom it shared several of these "Rivers of Empire."

On the Mexican side of the border, development was largely dependent on American schemes. There was virtually no settlement in the Mexicali Valley until the first and fateful diversion of the Colorado River. That famous experiment sent Colorado River water through Mexico and up north through the dry channel of the Alamo River to the Salton Sink, to which Charles Rockwood gave the revealing name, Imperial Valley. Worster underscored the importance of that name in his chapter on this first attempt of mastery over "the awesome Colorado." The chapter is "A Place Named Imperial" (Worster, 1985, 194–212). Farming in the Mexicali Valley followed in the wake of this new king.

For the first decades of the twentieth century, the bulk of the farmland in Mexico was not owned by Mexicans. It was a huge *latifundio*—a big farming estate—in the possession of a syndicate that included the owner of the *Los Angeles Times*, Harry Chandler. With 832,000 acres under his syndicate's control, he was one of the largest single landholders in all of Mexico. These lands only passed into Mexico's control in the late 1930s, when Lázaro Cárdenas, one of Mexico's most revered presidents, seized the properties as part of his national agrarian reform and land redistribution. This nationalizing move was part of the historical impetus for the United States to pursue negotiations leading to the 1944 water treaty, seeking more stability in the relations between the two countries.

From the perspective of Ernesto E. Coyro's monumental Mexican history of the almost century-long contest over the waters of the Colorado River, Mexico was always playing from behind, always with a weaker hand. Coyro wrote the definitive Mexican version of the negotiations in a long, detailed, two-volume history called *El Tratado entre México y Los Estados Unidos de America sobre Ríos Internacionales: Una Lucha Nacional de Noventa Años* (The treaty between Mexico and the United States of America concerning international rivers: A ninety-year struggle). It was published in 1976. Coyro was himself one of the principal Mexican negotiators with the United States for the water treaty. His history is abundantly documented and full of anecdotes. It allows Americans to understand the official Mexican view of the history of the negotiations with the United States over the 1944 water treaty.

Few Americans will have read this daunting history—it has not been translated from the Spanish and it has no index to its more than nine hundred pages. It is slow going. But it is indispensable for understanding the Mexican version of

a complicated history. And occasionally the prose lights up into passages of real fervor, real passion. These "best passages" often do not relate to the treaty itself. They almost always have to do with the way the United States enforced its will upon the river and upon Mexico.

Coyro's interpretation of the treaty process is that Mexico was not so much a victim as it was impotent to do better than it did. He is not interested in

*Mudflats stretch for miles at low tide at the end of the river.*

deriding the United States. He is interested in describing the forces that led to the treaty as signed. He calls what happened a "competencia disiqual," an unequal competition. The instruments in this unequal competition were twofold: the United States' national policy of expansionism and its superior technology.

After briefly lamenting what he colorfully describes as Mexico acting like a blind man stumbling through unknown lands in its negotiations, Coyro spends many passages detailing the ways in which the United States acted as an imperialistic and treacherous power—"su politica imperialista y perfida."

In the background of the treaty negotiations was a pressing Mexican consciousness of the loss of the entire Colorado River watershed to the United States after the War of 1848—a loss that is still viewed by many as the result of American expansionism. Mexico had never really possessed the territory it ceded to the United States. It had never occupied its vast wilds in the north. But Mexicans experienced a sense of profound injustice at the loss of these lands.

Mexicans feared the same fate for nearly all of northern Mexico, and this forms the context out of which Mexico operated in its dealings with the United States. Coyro details several events that compounded Mexican fears. One concerned the *filibusteros*—freebooters or buccaneers—private American citizens who invaded the northern Mexican states. These individuals' ambition was to set up an independent republic and then be annexed by the United States.

The most famous of the *filibusteros* was William Walker. In 1856, he and a group of men attacked La Paz, at the tip of the Baja peninsula. They soon retreated to Ensenada, where they took over the presidio and declared the "Republic of Baja California."

Coyro details invasions of various sorts up through the 1920s—and he claims that the U.S. government knew well of the attempts. The "mania anexionista bajacalifornia"—the mania to annex Baja California—included direct efforts simply to take over the Colorado River all the way to the Gulf of California. Coyro quotes a 1907 letter from the first director of the Reclamation Service (later renamed the Bureau of Reclamation), written after the disastrous floods of the Imperial Valley and urging that the delta problem be solved by acquisition: "if it could be possible to obtain a cession of the territory down to the head of the Gulf of Lower California, every difficulty would be easily saved. I find that some interested parties believe a purchase could be made" (Coyro 1975, 289).

Coyro claims that Mexicans were regularly threatened by force—including invasion—if Americans did not get their way with the waters of the Colorado River.

Technology was even more effective, though, in taking over the river. Created in 1902, the Reclamation Service was the decisive instrument of Mexico's submission to the United States' will. This understanding underlies Coyro's interpretation of the entire history:

Thus, the Reclamation Act of 16 June 1902 gave to the U.S. government the legal base and the power, such that the Reclamation Service constituted the technical and executive instrument, for control of the international rivers and for keeping Mexico from being able to exploit any river channel that came out of the United States. (Coyro 1975, 212)

As a result of Coyro's point of view, his history seems to narrate an almost fated outcome, an inevitable climax. Mexico could never compete with the federal resources that were poured into the arid West, he believes, much less with the technological achievements that led to the taming of the wild Colorado River.

Any illusions as to the relative outcome of the countries' water dispute were completely stripped away in 1928, when the United States passed the Boulder Canyon Act, a decisive historical turn according to Coyro. The critical element of that bill was the authorization of the Boulder Dam, later renamed Hoover Dam, a dramatic demonstration of "la disiqual carrera de aprovechamientos" on the river, the unequal course of damming the river (Coyro 1975, 527). Coyro summarizes the situation that Mexico faced:

> However, the passage of the Boulder Act, precisely in the middle of the treaty talks, necessarily imposed itself upon the Mexican Section as a radical change and demanded an immediate change in its own direction. There was no time to lose: the U.S. government was now limited in the negotiation in the amount of water at its disposal to give to Mexico from the stored water in the dams, because the water went to the states, made possible by federal funds; and the surplus, as it was easy to see and in fact turned out to be the case, would be the object of a fierce battle between California and Arizona, a battle which each day consolidated interests against any possible arrangement with our country. The technical and political situation was now very clear, and the threat to the interests of Mexico had turned imminent and grave. . . . (520)

The authorization and subsequent building of the dams left Mexico in an untenable position. From then on the "rude and unequal competition in hydraulic works and the extensions of the impoundments that took place at each step" signaled that the United States controlled the river (Coyro, 1975, 566). Certain historical factors finally made the United States eager to negotiate— notably Roosevelt's Good Neighbor Policy and the need for a secure ally along the southern border—and a need to settle any uncertainty that building Boulder Dam was in fact legal.

Viewed in this context, Mexico bargained well to get 10 percent of the Colorado River.

Coyro's interpretation is not a "merely" Mexican perspective of the treaty, and thus easily dismissible. The best U.S. historian of the 1944 water treaty, Norris Hundley, essentially agrees that the 1944 water treaty reflects the relative powers of the two countries during the negotiations. In a lucid and balanced narration of the long treaty negotiations, *Dividing the Waters: A Century of Controversy between the United States and Mexico,* Hundley underscores the role of U.S. power:

> Water is one of man's oldest and most important concerns, and some experts insist this concern prompted the location of the first civilizations. . . . Cynics have often described international law as being the will of the stronger nation, or the result of a power struggle among nations. To a great extent the cynics are right. International law lacks the substance that gives domestic law its meaning and force. (Hundley 1966, 17)

Hundley's title, *Dividing the Waters,* carries a biblical allusion, and highlights the almost religious fervor attached to desert water in the West. But this dividing of the waters has been a largely profane and greedy business. The current movement to attach a formal ecological minute to the 1944 water treaty might be understood as trying to balance power politics with an evolved sense of international environmental ethics.

But it would be naïve to see efforts to create a formal minute as a purely environmental altruism. Politics are still involved—only now there is an environmental constituency that simply did not exist in the 1940s. There is also a new view of border relations. Agriculture and the cities remain potent forces in the water culture of the desert. But they are no longer the only power brokers.

Hundley ends his history with praise for the 1944 treaty, saying it provided a successful mechanism for two rivals to live beside the river. His last sentences are a caveat that the treaty may remind one of Moses dividing the waters of the Red Sea. But when it comes to the Red Delta of the Colorado, the law of the river is not equivalent to the Ten Commandments: the law of the river is not a divine law set in stone. Rather the river of law evolves and changes to meet new historical needs, as Hundley writes:

> The United States and Mexico have made significant headway in the nearly century-long battle over their border streams, and, hopefully, their record of successes and failures will benefit other nations faced with similar problems. But any benefits that may have been achieved should not be marred by neglecting to solve newer points of controversy. (Hundley 1966, 188)

The delta in Mexico is one of these "newer points of controversy" that must now be solved.

The river has always been a kind of boundary between the two countries, literally for a certain stretch, and figuratively in the way it has been divided and the implications for wealth that have resulted. The river has been determinative in our geographies. In a real sense, altering the treaty enables us to modify these boundaries—to make them a little less hostile, a little more cooperative. In a more metaphorical but still real sense, altering the law of the river also means reconfiguring the geographies of our imaginations, bringing them a little closer to the real geographies of the lives in the delta.

*Kids play in an irrigation ditch early on a summer morning as pumps shower them with precious groundwater from the aquifer, supplementing treaty allocation.*

While the states, cities, and agricultural users mobilize in opposition to the delta, virtually every other expert has come to realize that the real issue is not

whether the delta in Mexico should get water. That question has been answered in the affirmative. If you love nature and wild creatures, you probably do not need the legal arguments to know where you stand.

The pressing questions about saving the delta are how much water does the delta need and where will it come from?

Steve Mumme is a professor of political science at Colorado State University. He has been a Fulbright Scholar in northern Mexico, studying water issues. Widely recognized as an expert on transboundary water issues between the United States and Mexico, he has a particular interest in the issues surrounding development and the environment. "The treaty has to be part of the delta solution," he told me in a conversation. "Mexico will be an ally to environmentalists on this. So among the specialists, the debate seems not to revolve so much around whether saving the Mexican reach of the delta is desirable, but rather around how you get the water."

Biologist and delta advocate Ed Glenn has provided a preliminary answer on amounts. As shown in chapters six and ten, he studied the cycles of regeneration in riparian habitats and in shrimp harvests after the floods of the 1980s and 1990s. He concluded that the current revival in the delta can be sustained by a mere 30,000 acre feet of freshwater in the river per year, plus a four-year cycle of "flood releases" in the amount of 260,000 acre feet. That amounts to about 100,000 acre feet per year, not including the water needed to sustain the Ciénega de Santa Clara.

That's less than 1 percent of the annual flow of the Colorado River. Not much, but in an overallocated river, hard to come by.

The Sonoran Institute, a nongovernmental organization that does community and environmental work along the U.S.-Mexico border, has helped sponsor a strategy for coming up with purchased water rights that could be dedicated to environmental restoration in the delta. In a study funded by the Packard Foundation, scientists and lawyers worked to identify specific sources of water for the delta. Their concern was not to solve the long-term ecological needs of the delta. Their concern has been to ensure that the delta does not dry up again after the Surplus Criteria dictated by the California 4.4 Plan go into effect and before a formal minute can guarantee a share of ecological water.

The group succeeded in identifying 30,000 acre feet of water for the delta's wetlands. It is an amount that satisfies Ed Glenn's estimate of the annual necessary flows. The study recommends that 15,000 acre feet come from marginal farmlands purchased on the Mexican side of the border, and that another 15,000 acre feet be diverted from irrigation runoff on the Arizona side.

Though a start, for complicated technical reasons this water might not be an adequate solution. "We can't afford to wait for a long-term solution," says Steve Cornelius, the director of the Sonoran Institute's desert ecoregion. Advocates believe that the proposed solution is at least a good down payment on water for the long-term delta needs.

There is a more troubling aspect to the search for water for the delta. It is the problem of water scarcity and overallocation itself. The question is how we as citizens can begin to face the resource limitations that, as a nation, we have largely ignored. Fundamental to this question is how long we can sustain an irrigation-based culture in the desert and how much growth desert cities can sustain under limited resources. These are large social policy questions.

The broad message of the most recent water wars in the U.S. and Mexican deserts is stark: We are stumbling blind into an era of constraints and limitations. The deepest question is not how to avoid paying our debts to nature. The deepest question is how we can begin to face the water facts and change our behavior.

Sandra Postel's excellent book, *Last Oasis: Facing Water Scarcity*, addresses what she describes as an emerging global crisis. The cities of the desert Southwest, like Phoenix, are living in an "illusion of plenty," she writes. Masking scarcity, she argues, has been the way most social policy has been conducted with regard to water. The harmful side effects of water overuse are hidden from the public. Consumption escalates, largely unconstrained, until a crisis leads to collapse. This has been the pattern in many cultures in history. Many wonder whether the American Southwest is following just such a well-trod historical path. Postel writes:

> Taking heed of water's limits, and learning to live within them, amounts to a major transformation in our relationship to freshwater. Historically, we have approached nature's water systems with a frontier philosophy, manipulating the water cycle to whatever degree engineering know-how would permit. Now, instead of continuously reaching out for more, we must begin to look within—within our regions, our communities, our homes, and ourselves—for ways to meet our needs while respecting water's life-sustaining functions. (Postel 1992, 23)

For Postel, the delta of the Colorado River is one of the central examples from around the world that demonstrates the need for a new ethic—and as important, new behaviors. *Scientific American* magazine in 2001 devoted a whole section to the politics and ethics of water scarcity as we enter the new century—a sign of water's growing national and global importance. The magazine issue includes an article by Postel on practical water-saving strategies for agriculture. But playing Cassandra to an ecological Trojan Horse like water scarcity is not the sort of message Americans like to hear. We would rather be told we can have it all, and have it right now, too.

But in the desert Southwest, the crisis has arrived. San Diego and Los Angeles were the harbingers of an inevitable future when they began to search for water efficiencies to meet the California 4.4 Plan. As Postel puts it, "Doing more with less is the first step. . . ." (Postel 1992, 23). The second section of Postel's book explores a number of practical mechanisms for using water more efficiently, some of them now being adopted by the southern California systems. These mechanisms include "thrifty irrigation" systems, like drip irrigation.

Postel also describes the importance of efficient toilets, which can save up to a third of the water usually used with each flush. She suggests eliminating leaks in water systems. And she advocates for raising the price of water, so that the price we pay reflects its true cost. "Studies in a number of countries, including Australia, Canada, Israel, and the United States, suggest that household water use drops 3–7 percent with a 10-percent increase in water prices" (Postel 1992, 155–6).

Global conservation issues typically come down to individual behavior. The average American uses 50 gallons of water per day: 19 gallons are used daily just in flushing the toilet, 15 in baths and hygiene, 8 in laundry, 7 in the kitchen, and 1 in housekeeping. What follows are just a few impressive water facts:

> ➤ A leaky toilet can waste up to 200 gallons of water daily.
> ➤ A leaky faucet wastes up to 20 gallons daily.
> ➤ A family of four typically consumes 700 gallons of water per week—
>   if they only take five-minute showers.
> ➤ Garden hoses discharge 6.5 gallons of water per minute at normal
>   pressure. Soaker hoses and trickle systems can reduce the amount of
>   water used for irrigation by 20 to 50 percent.

Conservation can directly affect the delta. Lisa Force is with the environmental and conservation organization, Living Rivers, in Tucson. As part of their campaign, "1 Percent for the Delta," she says that Americans could easily conserve enough water to save the Mexican delta.

For example, beef is very water-expensive to produce, yet cows are the major recipients of river water, as discussed in chapter two. Producing one pound of beef requires 2,600 gallons of water.

"If the 30 million people who receive Colorado River water in the United States each ate only one half-pound hamburger less each year," Lisa says, "we could save more than enough water to restore the delta."

She lists other ways that individuals can help conserve water for the Mexican delta:

> ➤ Turning off the water while brushing teeth saves 4–5 gallons.
> ➤ Flushing the toilet one fewer time per day saves 2–7 gallons.
> ➤ Aerators in kitchen faucets save 10 gallons per day for a family of four.

"If every American who depends on the river would save just three gallons of water a day," she says, "we could save over 97,000 acre feet of water." That's almost the 100,000 af needed for the Mexican delta.

Beyond the politics of the Mexican delta, perhaps the deeper message is this: We have a chance, a moment, to rethink not just our priorities, but our concept of water itself and our relationship to water.

⟨⟨

Water interests in the United States frequently take cover behind the complexity of issues they face in meeting demand—or behind the jealousies between the states. If Arizona saves water, they say, California will steal it. If we promise water to the delta, they say, Mexican farmers will tap it.

So why save water?

This is a mentality that may be, in the long run, self-defeating—an argument that could turn the water nations of the desert Southwest into pillars of sand as they drink themselves dry. That has been the fate of every other large-scale irrigation society in history, as Donald Worster warns in *Rivers of Empire*. Irrigation societies—water nations—construct themselves out of a defiance of nature, a defiance that comes back ultimately to destroy them. The two nations that grew up in the dry deserts of the North American continent, the United States and Mexico, follow, as Worster puts it, "a sharply alienating, intensely managerial relationship with nature" (Worster 1985, 5).

But we are changed in the process. First we come to serve the very devices we invent for dominating nature, in a circular logic that spirals out of control. And then nature turns back upon us, in accumulated salts and poisons in the water and soil, or in degraded fisheries, or in dams that have filled with silt, or in the number of endangered species that have multiplied in sad and frightful ways, or in the growing cities in the desert, where no more water is available. Worster calls the result an "ecological backlash," as sure and inevitable as history itself.

Such a fate was warned against by Percy Bysshe Shelley in the poem, "Ozymandias," often quoted by people worried about the trajectory of impossible growth in the Southwest. All the great irrigation societies, the hydraulic societies, that strove for domination of nature came to this barren end:

My name is Ozymandias, King of Kings:
Look on my works, ye Mighty, and despair!
Nothing besides remains. Round the decay
Of that colossal wreck, boundless and bare
The lone and level sands stretch far away.

The great monuments of water nations, reclaimed by deserts, become little more than "colossal wrecks." A disturbing question raised by our heavy use of water in the desert is, how long can such patterns of consumption last? We have thus far managed to live far removed from the realities of life in the desert; we have come to believe the illusion of a life of retired leisure in the desert. Air-conditioning and the hotel pool in the desert—they seem like the great symbols of a culture's triumph over nature. We have made the desert bloom and have turned hell green. We have found the golden life in the desert.

Among the dangers of our comforts and luxuries may be complacency. Perhaps the Mexican delta is the canary in the coal mine—the message that we cannot go on using the same excuses, pursuing our blinkered obsessions and the opportunity to act.

Sandra Postel's call for a new water ethic is a necessary corrective, offering new water behavior for a new ethic. Donald Worster also writes of an alternative. Historically speaking, he says, the enduring communities in desert environments were not constructed by hydraulic armies. "Where such communities endure," he writes, "the water flows and flows through history, as nature and human community join together in a single circle" (Worster 1985, 36).

We are a part of the natural system we have tried to control. It is inescapable.

The same might be said for our relationship with Mexico, which is in large measure refracted through the waters of the Colorado River. The problems along this waterway are part of a much larger national and even global water crisis. Nearly 40 percent of the world's people live in river basins shared by more than two countries: the Colorado River between the United States and Mexico, the Ganges River between India and Bangladesh, the Danube River between Czechoslovakia and Hungary, and the Mekong River between Thailand and Vietnam. Africa has fifty-seven rivers and lake basins shared by at least two nations. The most famous example of a world water crisis is in the Middle East.

Even in the United States, water problems and water shortages have spread now beyond the desert Southwest. The Pacific Northwest, the land of legendary rains, recently suffered a prolonged and expensive drought. Klamath Basin water issues between farmers and endangered salmon have made national headlines—precursor to coming water wars throughout the nation. Even in the wet East, around Lake Michigan, planners are predicting droughts and shortages. Businesses themselves are starting to imagine a global market for water.

Water scarcity throughout the world has led many thinkers to worry that water crises could grow into true global water wars. Peter Gleick, director of the Global Environment Program at the Pacific Institute for Studies in Development, Environment, and Security in Oakland, California, is a specialist in water and national security issues. He also has an essay in the 2001 *Scientific American* issue on water and security. In an another article called "Water and Conflict:

Freshwater Resources and International Security," he writes that in the coming century "water and water-supply systems are increasingly likely to be both objectives of military action and instruments of war as human populations grow, as improving standards of living increase the demands for fresh water, and as global climatic changes make water supply and demand more problematic and uncertain" (Gleick 1993, 79).

Water is for fighting, runs an old saying from the American West. History bears the scars of this truth. National security is increasingly a real factor along water boundaries. Just as the original 1944 water treaty was meant to serve as a model for the world, so does the current controversy over the water for the Mexican delta. The delta's predicament is giving us a model of new possibilities for changing our relationships with nature and with our near neighbors over water.

Despite a hopeful sense that the delta's time has come, there is a corresponding sense of gloom in the delta. And the situation has national security implications for the United States.

The new secretary of SEMARNAT, the environmental ministry in Mexico's new Fox government, is Victor Lichtinger. In a recent interview with Carlos Reyes, a writer for the newspaper *El Norte*, Lichtinger spoke frankly about water issues on the border as destabilizing the relationship between the United States and Mexico. "If we don't treat water carefully," he was quoted as saying, "it could become a true irritant in the relations between the United States and México, and could therefore be a destabilizing factor in North America" (Reyes, 2001).

Another Mexican who has been active in Colorado River water politics, Carlos Yruretagoyena, reacted to the Surplus Criteria in this frustrated outburst on the listserv for the Delta of the Colorado River:

> I really hope it creates a war between Mexico and the U.S. and that Mexico loses it and then the U.S. can take away what's left from the other war [referring to the war of 1848 that cost Mexico much of the American West] and that should take care of the water issues once and for all. Use water to its fullest. Produce as many crops as possible. Build a new aqueduct and screw Mexico's ecology just for the sake of keeping San Diego's golf courses greener. (Yruretagoyena, 2001)

The last time emotions in Mexico ran high like this, over salinity, the United States was forced to renegotiate the 1944 water treaty with a new Minute 242 to deliver clean water.

Water security issues came home to me most forcefully with Juan Butrón, the man who lives beside the Ciénega de Santa Clara and who loves it so deeply. He and I saw each other in the United States once, not in the Ciénega, but at the conference at the Mission Inn in Riverside, where Acting Deputy Secretary of the Department of the Interior David Hayes spoke. Juan had been given a small grant to attend the conference. The grant was given to him in a check in American dollars. He wanted to cash the check in the United States, where it would be much less of a hassle than if he had to wait for it to clear a Mexican bank when he returned home.

He asked if I could help him find a bank where he could cash the check as he speaks little English and had no car. Of course, I told him.

We visited a bank branch in a nearby shopping mall.

Juan lives in an *ejido* in Mexico that is very poor by American standards. It has little water, beyond the Ciénega, and few of the comforts like air-conditioning and paved streets that Americans demand. It barely has the necessities, much less the luxuries. No fountains. No swimming pools. For Americans, this shopping mall we visited in Riverside was like any other, upscale but not luxurious. For Juan, used to a very different scale of wealth, it was painful to see.

I asked how he perceived this display of material wealth.

His answer went a long way toward explaining why so many Mexicans stream across our borders, why our borders are so unstable.

"Yes, I have to say it hurts when I come here to the other side of the border and see how much wealth, how much water, they have here. We only want what is fair in the delta. It looks like we have to fight for it."

❧

A year after our cruise to the end of the river, I see biosphere reserve director José Campoy again in the delta. The Surplus Criteria have been adopted in the United States. Releases of water that reach the delta have dried up. I ask him if there is still water flowing in the delta from the Colorado.

"No more water," he says flatly.

Normally José is optimistic and accommodating. He continues to hope for a minute to the treaty that will deliver water across the boundary for ecological purposes; a new minute remains his most fervent hope for the ecological restoration of the delta. But the delta has been caught in the vise of the new water wars in the West. The delta is being squeezed between the two countries, Mexico and the United States, and José's job leaves him in the middle.

Even José has begun to reflect on the frustration many now feel as the United States grabs onto the Colorado River's excess flows in the Surplus Criteria.

"The Colorado River is a river, badly treated, but it's a river," he continues. "I would like to see the delta treated like it's part of the river. This is like Israel and Palestine fighting over the River Jordan. It's not a violent conflict, but I'm afraid there might be a bitter fight."

▶ *A soft haze of pink suffuses the air of a delta morning.*

# THIRTEEN

# A Sea and No Exit

SPECIES: Caribbean flamingo
PLACE: Salton Sea

THE DELTA IS A LANDSCAPE of extremes: of astonishingly high temperatures and almost no rain; of lush marshes that are among the most spectacular wetlands in North America and the driest deserts on the continent; of empty landscapes and some of the largest concentrations of people in the hemisphere. Nothing happens in moderation on the delta.

Even the natural beauty of the delta is hallucinogenic and at times, fiercely predatory. And pollution in the delta reaches horrifying levels. Ironically, the worst of the pollution in the delta happens not in abused and maligned Mexico, but in the United States.

The Salton Sea is the largest lake in California. It has also become famous as the most polluted lake on the continent—which is not strictly true. But it is a reputation this abused and manmade lake has not been able to shake. The lake is one of the strangest places in the country. It is located in the United States, it is a fantastic natural area with some of the worst environmental problems in the country, and it is a part of the delta. Some hate to admit it, since the Salton Sea seems to create even more problems for delta restoration.

But to understand the delta as a whole—and to work out a plan for its restoration in Mexico—we will probably have to figure out how to save the Salton Sea as well. For better or worse, the Salton Sea is one more example of the delta connected across borders.

Both beauty and pathology come together in a jolting clash of opposites this fall morning on the shores of the Salton Sea. Heat waves rise off the waters

◄ *The silhouette of a heron against the shimmering beauty of the Salton Sea illustrates why many hoped the beauty of the Salton Sea could be transformed into a real estate and vacation property boom land.*

▲ *Caribbean flamingos are startling and unexpected sights flying over the Salton Sea.*

and distort the distant view. I am staring through my binoculars in disbelief at three wavering, mirage-like smudges of pink. If they are what I think they are, they do not belong here. They are literally thousands of miles off course.

They are also one of the most beautiful birds in the world, a symbol of the perfect tropical life. But I cannot see them clearly, so I turn to Steve Miyamoto.

Steve is a biologist with the California State Department of Fish and Game and assistant manager of the Imperial State Wildlife Area on the southeast side of the Salton Sea. He has lived here with his family only a short time, helping to manage the area for wildlife—both for endangered creatures like Yuma clapper rails, which occur here in small numbers, and for popular hunted species, mostly ducks.

*A dead tree in the Salton Sea seems to symbolize the dying sea.*

261

We are out bird-watching together at the southeastern edge of the lake and he is showing me why the lake has a reputation as one of the best bird-watching locations in North America. I have been to the Salton Sea many times in the past. I know some of the best that it has to offer. But I was not prepared for what I was about to see: exotic pink birds at the edge of the salty shores.

"Are those Caribbean flamingos?" I ask.

Steve nods and smiles. "We get vagrant birds here from all over North and even South America. We've had as many as seventeen flamingos here at Salton Sea."

It is a bizarre moment for me. Only months before I had spent several weeks on the Yucatán Peninsula in Mexico, studying Caribbean flamingos (*Phoenicopterus ruber*) and writing about them for *Wildlife Conservation* magazine. I knew them well, including the fact that the Yucatan marked the northern limit of their range in the Western Hemisphere. They should not be here.

Steve is not quite sure what flamingos are doing in southern California. They could be escaped birds from a private collection. More likely, Steve says, they are migrants blown far off course from the Yucatán, where the species has made a recent comeback. Suddenly the whole hemisphere seems small and closely connected.

Wherever the birds are from, they have found a flamingo's paradise in the Salton Sea. Adapted to hypersaline conditions, flamingos have special glands that enable them to "sweat out" salts from waters that would be intolerable to humans. And as its name suggests, the Salton Sea is salty—and over the last century it has grown steadily more so.

In technical language, the Salton is called a "terminal sea," a geologically defining characteristic: the sea has no outlet. The waters that feed it have nowhere else to go, except via evaporation. All the agricultural salts from the surrounding Imperial Valley are sloughed into its shallow waters. As the waters evaporate, the salts have continued to increase, new salts are then added, and the cycle has turned vicious. The Salton Sea is now 25 percent more saline than the ocean. And climbing.

The stunning and unexpected beauty of flamingos and the unnatural pollution of the salty sea—they represent the two sides of the Salton Sea that are on a collision course. The sea combines some of the most important natural values on the West Coast with environmental problems—a strange coupling of pollution and high biological productivity—that has already resulted in horrifying epidemics for birds and health warnings for people. Yet you can see some of the rarest and most beautiful birds in North America here, not to mention some of the largest congregations of wintering waterfowl.

A lake in the desert is a magnet for birds, especially on the West Coast where the Salton Sea is now one of the key stop-over spots on the Pacific

Flyway. If you revel in the glory of birds, here is your place: congregations of ducks and avocets, stilts and herons. Some of the rarest inland birds on the continent are found here, with such euphonious names as fulvous whistling duck, or magnificent frigatebird, or blue-footed boobie. Just raise your binoculars and try to not to step on the dead fish, killed and washed up in the toxic waters. And hold your nose. Better brace yourself for the stench. The odor varies. Smells come from the fertilizers and pesticides on the 500,000 acres of surrounding farms. The sea itself also stinks from its tremendous biological activity, including decomposition of the dead fish.

The sea itself is 228 feet below sea level and it has no natural outlets. It is also huge, thirty-six miles long and between nine and fifteen miles wide. It is thirty feet deep and 230,000 acres in surface size. And for Colorado River water that feeds the surrounding farmlands, the Salton Sea is the end of the line, the catch basin for all the agricultural runoff in the valley. It is fed by 1.3 million acre feet of agricultural runoff. It already contains over 400 million tons of salts. Each year an additional 4 million tons of salts flow into the sea. It collects pesticides and fertilizers. Plus the most polluted river in North America, the New River flowing up from Mexicali with untreated sewage, dumps into the south end of the Salton Sea.

This sea with no exit sits in the desert, a festering brew of double trouble. In the late 1990s the federal government stepped in and funded $25 million worth of scientific studies and blue-ribbon panels. For all the money spent, what can or will be done about the Salton Sea is anything but clear. Several rounds of government-sponsored studies have produced various proposals to save the Salton Sea. Whether any is feasible is another question. All solutions are massively expensive—they could reach the billions of dollars—and most involve huge engineering fixes. Not everyone even believes this "mistake of man" should be saved, anyway.

Watching the flamingos along the shore of the Salton Sea, Steve and I decide to try to get closer. "This area is one of the best winter birding spots for all types of birds in the country, even the world," Steve says.

The flamingos are themselves an example of just how great a place

American avocets are one of the hundreds of species that make the Salton Sea one of the great birding places on the whole North American continent.

263

this poisoned paradise is for birds. The species list here has more than 380 birds, rivaling such hot spots in the United States as south Texas and south Florida. You can find some of the best rarities in the country only here—yellow-footed gulls, for example, fly up from Baja and are frequently seen here, and black skimmers have three of the five nesting places in the United States here on the sea. It is the only place where gull-billed terns nest inland.

Steve tells me there are about 12,000 white pelicans here now, and 2,000 brown pelicans. There is even a small population of Yuma clapper rails, maybe 200 birds. "You can come here and focus on the birds in the Imperial Valley and the ecology of the Colorado River and Gulf," he says. Every one of the birds mentioned above is found on both sides of the international boundary that cuts through the delta. Many of them will pick up and fly from one country in the delta to the other in a single day.

California has lost over 90 percent of its wetland habitat, which makes the Salton Sea even more precious for creatures. As one of the key stopovers in the Pacific Flyway, every year as many as 2.5 million ducks, geese, cormorants, shorebirds, and other birds who like water come here in the winter.

Caught in the squeeze of ecological worries, agricultural needs, and yes, even the long reach of thirsty cities, the Salton Sea survives in its own existential dilemma. Save it, at least in part for its stunning natural values. Or do nothing and watch it fester. Since the sea is, in fact, one of the greatest engineering blunders in history, it is a good question whether we should do anything but let it die its own slow, natural death.

The accidental creation of the Salton Sea is one of the most often-told tales of the Colorado delta. And for good reason. It was created out of a massive combination of greed and international deception that really opened up the whole delta on both sides of the border to settlement. Financial and engineering boondoggles transformed the old "Colorado Desert" into the most productive farmlands in the Western Hemisphere and left a set of ecological problems that nobody, one hundred years later, has yet figured out how to solve. The sea's creation is the origin of all the exploitation of the Colorado River and a microcosm of the delta's problems.

In 1849, a physician on his way to the California gold fields passed out from the heat and fell from his horse in the Mojave Desert. His name was Oliver Wozencraft. Crawling to a nearby *barranca*, or cliff, he gaped down into the vast and merciless desert of the Salton Sink, a huge desert bowl almost three hundred feet below sea level. It was desolate in the extreme. As he gazed upon the hallucinatory landscape, he imagined a utopia of farms and flowing water.

He was the first to imagine irrigating the Salton Sink with the Colorado River. He died having never realized his dream. And for decades after, men looked at the Colorado River and felt it was pouring "uselessly" into the Gulf of California, like a mockery to those who imagined the great wealth that the river could bring to these immensely fertile alluvial lands.

At the turn of the twentieth century, two men hooked up to make Wozencraft's hallucination in the desert a capitalist's dream. Charles Rockwood, an engineer, and George Chaffey, a pioneer of irrigation projects in southern California and Australia, formed the California Development Company. Their scheme was to turn a portion of the Colorado River away from its channel and bring it into the Salton Sink. Since huge dunes blocked the way through California, they finagled the right-of-way through Mexico with a Mexican dummy company, and routed a sixty-mile canal through the Alamo River and into the Salton Sink. The whole complex scheme still galls Mexicans, though it was a century ago.

In May 1901, the twentieth century opened in the West with the Colorado River churning into the desert. Rockwood and Chaffey renamed the "sink" the "Imperial Valley," testimony to their grandiose dreams. The men flourished—for awhile. By 1904 they had sold land and water rights to seven thousand alfalfa-growing farmers, who poured into the desert along with the water from the Colorado River. Then all hell broke loose.

It is a famous story—how Rockwood and Chaffey underestimated the flows of the river, how they thought they could mess with the river without properly appreciating its power or taking even normal precautions. In 1905, the first cut in the river silted up. They breached the riverbank a bit farther south, in Mexico. But they could not afford a head gate, so they took a chance that the river would remain low and tranquil, as it had been the first four years. The river made them pay. Big time. In February 1905, floods of geological proportions came roaring out of the Rockies and tore the breach wide open. Within days, the channel to the Imperial Valley was sixty feet wide. After the river scoured out the channel to a quarter of a mile wide, the entire river began seething through it.

The waters were unstoppable. Engineers labored sixteen months and spent $3 million to close the intake in the bank, but not before the river had created the Salton Sea out of the desert—seventy-two feet deep and now, after a nearly a century of evaporation, still 380 square miles in size. The Salton Sea is the legacy of the original and fateful incursion of irrigation into the delta—a legacy of greed and, subsequently, of neglect.

The Colorado River had often turned on its own when it silted up its channels to the sea, flip-flopping around the delta. The river had several times flowed into the sink and created great prehistoric desert lakes. The most recent one, Lake Cahuilla, probably dried up as recently as 1580. That lake left a bathtub ring around the mountains surrounding the sink as it slowly evaporated back into the salty oblivion of this desert.

Everyone expected the Salton Sea to evaporate under the desert sun.

In one of the strangest twists to the story, the federal government actually gave ten thousand acres of land under the Salton Sea in trust to the Torres-Martinez Desert Cahuilla Indians. It was land they had owned before the 1905 floods. The tribe was to take possession after the fourteen years it was calculated that it would take the sea to evaporate. The federal government only recently compensated the tribe for the lost lands.

The sea should have evaporated. Would have evaporated. But it did not.

Why? Farmers were undaunted by the floods and in fact redoubled their efforts to bring the Colorado to heel—culminating in the 1935 opening of Hoover Dam and the All-American Canal that brought well-disciplined water from the river through American land to the Imperial Valley. A complex array of water rights guarantees delivery of nearly 25 percent of the annual flow of the Colorado River to the Imperial Irrigation District and the Coachella Valley Water District—federally assured and subsidized—the single biggest users on the river. Ever since, agriculture has grown into the defining feature of the Salton—500,000 acres and a $1.4 billion-per-year empire of its own.

The Imperial Irrigation District and the Coachella Water District receive twice as much water from the Colorado River as the entire allocation to Mexico—about 3.4 maf. All this water is dedicated to agriculture or to water-transfer sales.

Agricultural interests quickly discovered that the Salton Sea was a cheap place to dump their runoff water. In the 1920s the Salton Sea was designated as the Imperial Valley's catch basin. To enable agriculture in the valley to flush their soils and leach out the salts, one of the most complex hydraulic engineering projects in the world was carried out in the valley. In addition to almost 1,700 miles of canals, there are more than 32,000 miles of tiles under the ground—tiles that send the water off the farms. With its load of salts and pesticides and fertilizers, all this water is shunted into the Salton Sea.

The sea became a cheap catchment basin for 1.3 million acre feet of agricultural runoff every year. Since the sea has no outlet, these polluted waters accumulate and thicken as the water evaporates at the rate of about five feet per year. It only rains two and a half inches in the area each year but instead of disappearing, the sea has actually risen over time as these wastewaters pour in. At the same time, all the salts, fertilizers and pesticides in the agricultural runoff waters have been growing more and more concentrated.

Today the Salton Sea is a sink of agricultural nutrients, surrounded by an empire of agriculture. Salinity of the sea has climbed to forty-four parts per thousand. The ocean averages about thirty-five parts per thousand. Nitrogens and phosphates from fertilizers promote a superabundance of phytoplankton, which turn the water a reddish brown and give off a wretched stink when they die. Pesticides are piling up. Selenium, a mysterious sulfur-like element, is at toxic levels.

The eutrophication (algae blooms) from the nutrient overload comes not just from agriculture, but from cities. The New River flows north out of Mexico. The New River, which makes Mexico's Río Hardy look clean by comparison, glugs into the Salton Sea in a brown sludge of raw sewage, industrial and slaughterhouse wastes, toilet paper, dead dogs, and human trash. It simply pours untreated into the sea. Natural processes clean up much of the contamination from Mexico, but it is still an unsettling sight to see, all that brown sewage surging into the sea.

On bad days you can smell the stench from the sea, and the cattle feedlots and fertilizers in the valley, all the way to Palm Springs.

But nobody much cared. The blue sea beckoned investors and dreamers. Real estate developers devised schemes to lure people into the desert to buy lots and water-ski on the lake. Places like "Desert Shores" and "Salton City" and "Salton Sea Beach" sprang up, each christened in the euphemisms of real estate semantics. But when birds started to die and the sea levels flooded lots, the mini-booms went bust and most people abandoned the sea. It has become a poor man's polluted paradise, where you can find low-rent and run-down RV parks and empty motels.

The neglect of the sea came to a spectacular and terrifying end in the early 1990s. Throughout the decade a series of apocalyptic die-offs of birds and fish turned national attention to the troubled sea. In 1992, the first large-scale die-off of birds began—called an epizootic. Eared grebes, a beautiful ducklike bird that dives for small fish, began to be seen staggering along the shorelines, disoriented and falling down. Gulls flew in and tore the flesh off the living birds. The death toll rose to 150,000 eared grebes in just a few months, their carcasses littering the shores. To this day, biologists are still not certain what caused the epizootic in the grebes. However, necropsies showed the grebes had three times the acceptable amount of selenium in their flesh. A mysterious element, selenium is necessary for human survival, but poisonous at elevated levels.

In 1996, more than 1,200 brown pelicans—an endangered species—died of avian botulism, together with almost 19,000 waterfowl and shorebirds from 63 other species.

In 1997, 10,000 birds died from 51 species.

In 1998, another 17,000 birds from 71 species floated dead on the waters and cluttered the beaches of this one-time paradise, mostly from Newcastle's disease and avian cholera.

At the same time, millions of fish started dying as well. When the sea was first formed, trout tumbled in with fresh river waters. These fish are long since gone. But the California State Department of Fish and Game worked with the growing salinity to keep this a poor man's fishing haven alive. By the 1950s the sea's salinity equaled that of the ocean. State biologists traveled to the Gulf of California and netted whatever they could catch. They dumped the fish in the sea and the stocks have become extremely popular—especially orangemouth corvina (*Cynoscion xanthulus*) and gulf croaker (*Bairdiella icistia*).

By the 1970s the Salton Sea was one of California's most popular sport fisheries. Then in 1986, the state issued warnings about eating the fish. Suddenly newspapers and magazines were running stories calling the Salton Sea a toilet bowl, an "environmental Chernobyl," a "fly infested sinkhole . . . of botulism-carrying maggots, crazed fish, and drowning birds." Fishermen and other tourists got the message. Fishermen have declined by about 50 percent on the sea, which turned into a body of water in perpetual crisis.

Now fish lie rotting on the shores, offering as gross and eloquent a warning as you could need. Strangely, all the pollution actually attracts the birds. They feed on the phytoplankton and the dead fish and all the other stuff that grows on the agricultural nutrients in the waters. There is nothing more disturbing, though, than watching shorebirds pecking away at the dead tilapia that are as thick as seaweed at the water's edge.

Bird-watchers have flocked to the sea in pursuit of the birds, making their quest for sightings one of the new growth industries of the Salton Sea. There is great birding, no question. But you have to focus on the birds all around and try not to think about what is in the water.

⁂

From a distance, the Salton Sea glimmers under the desert sun in a sparkling, beckoning blue—another prime piece of real estate in the geography of the California dream. The mountains on either side—the Chocolate and the Anza Borregos—loom through a soft veil of blue air and make you forget where you are in this troubled sea.

The Salton Sea is only about forty-five miles from the border with Mexicali. Yuma is barely an hour away. From the north end of the sea, the rich man's mecca of Palm Springs is only forty-five miles. And San Diego is only 120 miles to the coast. The sea is surrounded by the burgeoning world of the arid Southwest—right smack in its center. That only makes its future more dubious. In the geography of southern California's growth, the Salton Sea is located nowhere near any viable solution.

Like this entire scorched and parched region of America, everything with respect to the sea hinges on water. And even though this is a huge body of water, it is more mirage than reality. Fixing the sea is going to involve finding or fixing water, and so far the solutions that have been proposed are big-engineering fixes that look more like the past than the future.

The future for the Salton Sea has gotten more uncertain as the cities of San Diego and Los Angeles scramble for new sources of water. In the water transfers that secured more than 0.4 million acre feet of water for the cities in the California 4.4 Plan, 200,000 acre feet have been scheduled for purchase from the Imperial Irrigation District. Water is becoming more valuable to agricultural interests as a commodity than as a resource for farms. It is increasingly more profitable for farmers to sell their subsidized water to cities than to put in crops. Until now the sea survived, barely, on water that nobody else wanted. That tide has changed as water gets more valuable. That means less water is available for runoff to the sea.

The polluted New River is on the verge of being cleaned up and Mexico is not likely to send clean water to the United States and the Salton Sea—in which case, there goes another 22,000 acre feet per year. The sea is caught in a vise between the old economy of big agriculture and the new economy of big cities.

The less water that flows into the Salton Sea, the faster the rates of growth of the salinity. The fish, even the introduced fish from the sea, are in imminent danger of being salted out. When the salinity reaches fifty parts per thousand—expected in the near future if the current rates of inflow are maintained, and much faster if they are not—most of the resident fish will die out because they will no longer be able to reproduce. When the fish die, most birds will leave too.

Salinity, deadly eutrophication due to fertilizers and organic waste, pesticides, loss of water—the list of threats is formidable. Perhaps impossible. The Salton Sea is on a fast track to become a hypersaline "Dead Sea." But in 1998, Representative Sonny Bono, a Republican from Palm Springs, died in a skiing accident. He had grown up waterskiing at the Salton Sea. He loved the sea for sentimental, nostalgic reasons and had made it a pet project. His widow assumed the cause on his behalf. The national wildlife refuge at the south end was given the cumbersome, unrememberable and nearly unspeakable name of Sonny Bono Salton Sea National Wildlife Refuge.

That same year, another massive pelican die-off jolted Congress into action. In two years $25 million of state and federal money got pledged to the Salton Sea Authority and the Bureau of Reclamation for studies on possible clean-up scenarios. The results—some fifty-four alternatives—came down to two types of engineering models. The goals of the plans were all aimed at reducing salinity and maintaining sea levels. They did not address pollution.

By 2001 the Salton Sea Authority had developed a set of alternatives that make saving the sea feasible—if they can be coordinated with the major southern California water transfers. The major engineering fixes for the Salton Sea have all involved finding "outlets" for the Salton Sea—places where the heavily saline water of the sea can be directed and then evaporated.

Early in the process of looking for an engineering solution to the sea, people imagined such ambitious schemes as pumping some of the heavily saline and polluted water in the Salton Sea to, say, the Gulf of California in Mexico, or even to the Laguna Salada. After fierce protests from Mexicans, those ideas have been abandoned. In their place are two proposals now to build "outlets" in the sea itself, or nearby it.

In January 2000 the Salton Sea Authority issued a draft environmental impact report with five alternatives, all of them calling for evaporation ponds or towers. Water would be moved through a series of evaporation ponds, much like those used by private industry in producing sea salt. There are a lot of components to these plans, but they follow essentially two models. One version of these proposals is solar powered only—the water evaporates under the fierce sun. The second version uses a mechanical system to blow water into the air, enhancing and speeding evaporation and using less land than the system of solar evaporation ponds.

According to Tom Kirk, director of the Salton Sea Authority, who is responsible for developing a plan for the sea, either of the basic models is relatively modest as major environmental recovery programs go: about $300 to 400 million. Compare that, he says, to several billion dollars needed for the Everglades.

These prices, however, assume that the water transfers under the California 4.4 Plan to San Diego do not take place. If there is less water coming into the Salton Sea from agricultural runoff, less inflow, "the prices of lowering the salinity level grow drastically. With less water running into the sea after the transfers, the Salton Sea Authority estimates that solar evaporation ponds would have to pull an additional five million to seven million tons of salt out of the sea—at a cost of an additional $200–$300 million. This operation would be the equivalent of pulling a two-mile-long freight train full of salt, every day, from the sea.

"Water transfers put the cost [of saving the sea] over $1 billion," Tom says, "and may not be practical."

Not everyone agrees that anything should be done. Delta advocate Ed Glenn is one of those people. Without the engineering fixes, the sea would not necessarily "die." It would turn into a hypersaline lake, like Mono Lake to the north.

"It's not clear to me and many others that the Salton Sea is essential," says Ed. He suggests that the Salton Sea Authority restore and recreate habitat in the rivers that feed into the sea. Then let the sea run its course. "Those hypersaline ecosystems tend to be more stable. That's the way nature wants it to be, anyway."

What seems to be emerging, however, is a growing realization that the Salton Sea and the Mexican delta need to be considered jointly. For a long time, the two issues seemed to be competing against each other: you could save the delta, or you could save the sea, but you could not save both. But increasingly it appears that neither will be saved until both issues are resolved.

"Two years ago, even," Tom Kirk says, "there was a lot of jealousy among advocates for the Mexican delta for the political and financial wherewithal of the Salton Sea. And the Salton Sea was jealous of the attention the delta gets from environmentalists. But we can't ignore either, especially if you're concerned about the Pacific Flyway for birds. You've got to look at the whole ecosystem."

The feeling has been, it will be hard to do one of these projects so how can we possibly do both? Yet according to Tom, the two parts of the delta actually require very different kinds of attention. "If you look at their needs," he says, "they are generally different." The Salton Sea has no outlet, and it needs money to solve its problems. What it does not need is water and a better image. The delta in Mexico, however, has an outlet. It needs legal changes, scientific study, and a budget of water.

"I'm not afraid to think about extra-territorial issues," Tom says. "I'd like us to solve the institutional impediments to cross-border water transfers. Mexico is the underdog and the environmental community loves an underdog. But the abundance of bird life and biodiversity in the Salton Sea is even greater than that of the Mexican delta."

Nobody knows when we might begin cleaning up the Salton Sea. It has been an issue for at least forty years already. Tom believes the next year or two are likely to be critical—it's do or die time for the Salton Sea. He is hopeful, though, that a solution is now possible; that the two portions of this strange and ironical delta—the Salton Sea in the United States and the delta down in Mexico—can once again be thought of as a whole, despite the boundary between them and their very different environmental circumstances.

◖

Steve Miyamoto and I drive out on a spit in the south end of the sea to try to get closer to the flamingos.

"In an airboat you can get really close to them," he said.

We are not far from where the Alamo River flows into the Salton Sea—not nearly as ugly a mess as the New River, but not comforting either. No sign marks its entrance, identifies it. It pours into the sea anonymously, like a secret shame. But the inflow is a congregating place for birds. The organic wastes in the river make the waters fertile for fish and bugs and birds. The river pours out into the sea in a plume of reddish brown.

"It's kind of like a big cesspool," says Steve.

Close in on the spit is a beautiful flock of black skimmers, resting in their streamlined profile on the sand. Marbled godwits feed on dead fish on the shoreline. Avocets and stilts, blue herons and great egrets fish in the shallows. In late

Caribbean flamingos are often seen on the Salton Sea. Flamingos like these on the Yucatán Peninsula of Mexico are adapted to hypersaline conditions and may well inherit the increasingly salty sea if something radical is not soon done to save it.

July, there can be as many as 20,000 nesting herons and egrets and ibis in the area. I am watching a nearby flock of about 200 long-billed curlews preen themselves with their long beaks. Beyond them in a bay is an enormous flock of white pelicans. The whole blue bay is dotted with white. "The biggest flock of white pelicans I've seen in my life," Steve says.

Steve got his degree in wildlife management from Humboldt State University. He grew up in San Francisco and has only lived a brief time in the desert. He is a northern California person. I ask him how he likes this area. "I know what's in the desert, now," he tells me. "There are no trout in the desert. So it's not suited for me." Steve's job is to manage the habitat for birds, which he clearly enjoys and takes pride in. He handles the wetlands almost as if they were an agricultural crop. Recently, he explains, he has been flooding fields for waterfowl that will spend the winter.

"It helps lots of other species, too," he says.

But it is also clear that Steve does not really feel like he belongs here. But then, when it comes to the Salton Sea, nothing seems to "really" belong here anymore. It is a paradoxical natural habitat, like so much throughout the delta—the product of human blunders on a monumental scale, heavily abused, and yet one of the most fertile places for birds in North America.

I ask Steve what he thinks the answer to the Salton Sea will be. He hesitates. "It's an interesting body of water," he says. "It's got lots of history. It took us a century to foul it up. It'll probably take that long to fix it."

As we drive out on the spit, he points out Mullet Island. In the 1940s, this whole area was a popular vacation destination. "A guy named Robinson put a restaurant, a club, and a museum out there. The sea was lower so a road went out."

Now a shallow sheet of water lies over the road, taken over by the birds. The shapes of the bright pink flamingos dance on the heat waves in these shallows. They slowly come into focus, unmistakable. Caribbean flamingos are the biggest flamingos in the Western Hemisphere, the brightest colored flamingos of any of the six species of flamingos in the world. In the Yucatán, you can see these flamingos in some of the most astonishingly saline waters. They even stalk on their long legs through the most salty evaporation ponds of sea-salt companies.

The birds seem so completely exotic in this scene. Few creatures can survive in hypersaline waters. But flamingos can, feeding on little brine shrimp. That these birds should be so beautiful and yet live in water so salty it could eat your skin has always struck me as one of nature's beautiful contradictions. These flamingos do not belong here. But then, neither does the Salton Sea itself.

And if nothing is done soon, the out-of-place and spectacular flamingos are poised to inherit a landscape of extremes and a sea of salt.

# Making the Delta Real

SPECIES: Eared grebe
PLACE: El Doctor Wetlands in the Mexican delta

WHEN I VISITED the Mexican delta of the Colorado River for the first time, I flew to Mexicali from Mexico City, where I was living for the year. While in Mexico, I had been doing preliminary reading on the delta and I carried with me on my trip several of the notions I had gleaned from that reading. Before I even arrived in the Mexican delta, I already had several defining images and metaphors in mind. And I had never even set eyes on or foot in the place.

I imagined that it was the hottest and driest and most forbidding place on the North American continent, and as a result, would be largely hostile to humans. I knew as well that the delta was in what had been an empty desert, but that it had been transformed by settlers and engineers into a fast-growing part of Mexico. I knew as well that the entire delta—United States and Mexico—was now one of the most productive farming areas in the entire Western Hemisphere. Most of all, I knew that the delta in Mexico had been left for dead after the development of the dams upriver, all but one of them on the U.S. side of the border.

None of these images is wrong. But they could have blinded me to other features of the delta—the very ones that have become more powerful for me. My initial thoughts about the delta were the common and stereotyped images that both Americans and Mexicans have come to accept as defining the delta of the Colorado River. The pictures are a mix of truth and falsehood, of fact and fantasy. The dominant myth of the delta in our time—the time of dams—is just such a mixture.

*◄ The dock and boat on the Ciénega make it clear that this place is a home to people and creatures. In this desert delta, people and nature come together through water.*

*▲ The eared grebe approaches close. Locals in Mexico call this bird "pato marijuano," the marijuana bird, for its brilliant red eye.*

275

One of the realities I encountered almost immediately on my first trip to the Mexican delta was that is not completely dried out. It is abused. It is mutilated. It is largely abandoned in its wildness. But get below the farm levees, and you can find a lot of living wetlands. Even a day's excursion onto the Ciénega de Santa Clara will disabuse anybody of rumors of the delta's death.

I began to feel, very quickly, that the image of the dead delta may have been a little too comfortable, even self-serving, for a culture eager to enjoy the material advantages conferred by the great dams on the Colorado River. If the delta is already dead, why would we have to worry about changing our behaviors or think about restoring it?

Before the dams were built, the delta was viewed as uninhabitable and dead. Now that image of the delta has a very modern twist—now it is dead for lack of water and so those portions of it not given over already to agriculture can be dismissed and forgotten. Yet this image of the dead and barren delta is actually an extension of the image of this desert part of the continent that has prevailed for the almost five centuries since the delta was first explored by Europeans. What you begin to realize is the delta has had to be repeatedly "rediscovered"—first by the Spanish explorers of the 1530s, later by Spanish missionaries of the 1700s, then the gold rushers and steamboat captains of the nineteenth century, and most decisively by the settlers and farmers after the Colorado River was diverted and dammed in the twentieth century.

I am not sure that it would be an exaggeration to say that the delta has been one of the most ignored parts of the continent, largely invisible to us, shunned as unsuitable for habitation. It has often been described as an illusionary wasteland: a place of mirages and hallucinations; a fantastic place right down to the myths of California as a island inhabited by Amazon women, a notion perpetuated from the very instant of "discovery." Yet many of the fantasies to be encountered in the Mexican delta are the ones we have projected onto it.

Now we find ourselves in another wave of rediscovery in the delta—and this time what we have discovered is a whole new set of social and cultural values connected to desert climates and desert rivers. For this new discovery of the delta to take hold, a new image for the place will have to prevail. I came to this realization one morning while I was canoeing on the Ciénega de Santa Clara with Juan Butrón, the local protector of the great marsh.

Juan and I were paddling back from an early morning on the Ciénega de Santa Clara. We had had the Ciénega's lagoons and cattails entirely to ourselves that morning. No one else was on the waters. We had been at my single favorite place in the delta, the delta within the delta—the muddy ooze at the mouth of the Welton–Mohawk Canal. I have come to think of this spot as the living epicenter of the wild Mexican delta. We had seen coyotes and raccoons, watched brown pelicans and barn owls in the first light. I was filled with a sense of how much I had come to love this marsh and all the creatures that inhabit it over my many visits, and with how much affection I had come to feel for Juan and his family, as well.

Juan and I were laughing and joking. At one point, he said to me that he thought I had come to know the Ciénega very well. So well, he said, that I had become a *Ciénegero*. It was a wonderful compliment, a moment of connection. Juan made the word up, on the spot I think. But its meaning is easy to understand for anyone used to the idioms of Spanish. It means something beyond simply knowing about the Ciénega and its creatures. Juan had said I was a "person of the Ciénega."

Beyond the affection it aroused in me, I realized something else at that moment. The Spanish language makes a distinction between two kinds of knowing—and they are notoriously difficult for native English speakers to come to master. In English, we have only one verb that means "to know." It covers many different meanings, which in Spanish are easy to confuse and confound. One verb, *saber*, means to know a fact. Another verb, *conocer*, means to be familiar with something in a more intimate way, a more experiential way. You *saber* (know) where a place is located or what might be found there; but you *conocer* (know) a place from having visited it often or having lived there.

Juan's compliment to me was a nod to the *conocer* form of knowing that I had acquired by many visits, by getting out of the canoe and tromping through mud, by finding rails and dowitchers and pelicans and pupfish.

In the rush of feeling of that moment of affirmation, I realized something about the way we have used language to abuse the delta. We have called it dead. We have buried it under this metaphor as a way to put it from our minds. All the while, people and creatures have been living here, rebounding in numbers and exploring ways of living that put the wilderness along the U.S. portion of the lower Colorado River to shame. In that intimate gesture by Juan, a growing sense I had been developing suddenly coalesced for me in a new metaphor for the delta.

This is not a dead place. The Mexican delta is a home.

The very word ecology derives from the Greek *eco* which also means home. Ecology is the study of "home," and for humans this means "home" in the context of all the other inhabitants of the earth. The delta is home to a long list of birds and mammals and reptiles and fish—creatures. It is home to endemic sea grasses and salicornia, tules and bulrushes, cottonwood and mesquite forests—plants. It is home to Cucapá Indians and *ejiditarios* and unrecognized water users like those along Río Hardy—people. It is home to Juan and his son Katán, to Don Onesimo and his daughter Monica and her daughter, and to Jesús and Javier Mosqueda, as well as many others.

Many of the creatures are endangered, as is much of the eroding delta itself. Still, creatures like the flat-tailed horned lizard prosper here in Mexico in the still "empty" spaces of the desert around all the water along the river channel. The endangered people of the delta—the Cucapá and the small farmers and the ecotourism operators—have felt a new rush of hope as waters returned to the river following the floods of the past twenty years. But all these creatures and plants and people who call the delta home have been left out of the legal consensus upriver that divides and allots the Colorado's water. They are the other delta. The forgotten delta. The delta that was left out.

Perhaps now their time has come. The question is how to make them real to us.

Beyond the legal battles over the Endangered Species Act, or the political negotiations for a new minute to the 1944 water treaty, or even the scientific efforts to document the abundant life that was and is in the delta, I have come to believe that this question of language—of what words we use for the delta, what metaphors we apply—is fundamental. Through nearly five centuries, people of

*Juan Butrón calls the delta his "terreno nativo," his native land.*

European descent have viewed the delta as a kind of theater of conquest of one kind or another—a place to conquer in search of gold and pearls, for the first Spanish conquistadors; a place to convert human souls, for the Spanish missionaries; or a place to dam and turn to farming gold—the white gold of cotton, the green gold of alfalfa, and now the liquid gold of water itself—for modern water users.

We have made the delta into an "other," an alien place, not seeing all the while that it was already home to people and creatures. Through our metaphors, we may have imposed ourselves upon the physical landscape, but we have not yet learned to live there. We can survive this arid land with air-conditioning and cable TV and swimming pools—all dependent upon the dams on the river. But people like Juan and Monica, who have pursued a different notion of home in the desert, live modest and sustainable lives. They are trying not simply to live here, extracting and consuming. They want to inhabit the delta.

The delta is not simply a question of culture versus nature. It is a matter of what kind of culture we want to have. And that is, finally, a question of language—of the words that shape how we think about the world we inhabit, and of how those words shape our action. I came to the delta with one set of words in my mind and slowly had to set those aside, or qualify them, and learn another language for the place. I was tutored by scientists like Ed Glenn and Osvel Hinojosa, who have documented the new vitality of wetlands and birds in the delta; also by Karl Flessa and Salvador Galindo-Bect, who have demonstrated clearly that the dams upriver and American management of the river are directly responsible for the devastation in the delta. Don Onesimo Gonzalez and Monica Gonzalez and Juan Butrón—these people and their families gave me a different kind of tutoring. Perhaps more than any, the stories of these people touched my heart because they are living the delta. Not just in the delta: Their lives are the delta.

The delta in Mexico will not be restored unless the United States finds water to help it. After so much travel in the delta, especially on the Mexican side, and after so much research through reading and interviews, arguments that Mexicans are on their own in saving the delta seem to me almost cruel and willful. The ethical burden is largely on the United States, with its many dams. Mexico has only one dam, and that for the delivery of their treaty water. The United States uses the lion's share of the water in the river.

The two countries' destinies have always been conjoined in the delta, and this is true now more than ever. What is most moving to me in all this is that the most profound challenge posed by the delta is an ethical one. Yet it is not a call to anything like a radical re-visioning of modern culture in the desert—although our ethical deeds may eventually lead us there.

The Mexican delta is only asking for about 1 percent of the river's annual flow.

In his beautiful book on ranching and farming in the West, the writer William Kittredge speaks frequently about the importance of stories and language in shaping our ability to renew ourselves and the land itself. He grew up on the largest farm in eastern Oregon, and *Hole in the Sky* chronicles his personal struggle to find new language, new stories, to guide him into a new and more authentic relationship with the earth. "What I am looking for," he writes poignantly, "is a set of stories to inhabit, all I can know, a place to care about" (Kittredge 1992, 9–10).

It is a task we all face—as water becomes more contested and more scarce, as we spread and expand and grow in numbers. Perhaps it is the fundamental environmental task of our generation—to develop new notions of the world as home, "a place to care about." This means creating new stories that make a place like the Mexican delta real.

Making the delta real also means finding new ways to speak about it—"knowing" it in richer and more *conocer* ways. The early conquistadors like Cortés and his explorers imagined they knew the delta even before they set foot on it, from the chivalric literature they consumed. Modern Americans learned to see and "know" the delta largely through the lens of the simulated river of law which, projected onto the real river and its delta, transformed both. We want now to be sure that our words reflect and preserve the integrity of the place. We want to be sure that our words are rooted in the delta itself.

As is often the case, great ideas take hold in our lives through specific moments. I would like to end this book with one of the most exhilarating and moving moments I had in the delta—one that called me to recognize the delta's intrinsic right to be, to exist, to survive.

It happened during one of my stays in the Ejido Luis Encinas Johnson with Juan Butrón. One evening, he and I joined two biologists working for the biosphere reserve. María Jesús Martínez Contrerar is chief of projects, and she came with Miriam Lara Flores, a young volunteer biologist, the same woman who was with us when we captured migrating songbirds in the cottonwood-willow forest beside the river. Both of these biologists had helped as well with the Yuma clapper rail surveys.

On this night, the women are going to one of the natural springs on the eastern side of the delta, south of the Ciénega. They want to capture *pupos*—desert pupfish—in the ponds on the mudflats. Juan and I join them.

It is a chance to see some of the restoration projects that the biosphere reserve, working with many organizations in both the United States and Mexico, has undertaken in specific locations. One such project is located very

*Miriam Lara Flores (left) and Maria Jesús Martínez Contrerar (right), both biologists with the biosphere reserve, capture desert pupfish in the El Doctor freshwater springs. Note the fence to protect this pond from cattle.*

near where we are heading. We drive down the highway and turn off just south of a military checkpoint, romp across sand dunes, and drop down close to the mudflats. There we leave the car and walk through an area of springs known as El Doctor.

The first freshwater spring we encounter is a huge oasis surrounded by cottonwoods and willows. The water rises up from the aquifer, through subterranean faults, and forms a gorgeous deep desert spring. In the trees, we see two southwestern willow flycatchers, the endangered bird that migrates through the delta, using water holes just such as this.

María shows me an area of tall grasses and reeds that has been fenced off. Funded by private sources, the fence lets the wetlands grow up again by keeping cows out. The cows have free run of the land in this wet part of the delta. They eat everything. Saving spots for nesting birds and surviving fish requires such initiatives, and this is one of only several restoration projects in the delta— on-the-ground efforts that are valuable in their own right, and that show what can be done everywhere down here if the delta gets water.

Most of these projects have been undertaken by nongovernmental agencies working together on both sides of the border, like Pronatura in Mexico and the Sonoran Institute in Arizona. But they amount to people working together out of care, almost always working for little money. As I look at this fenced pond, I think of an earlier conversation with Mark Briggs, a specialist in wetland restoration with the Sonoran Institute. "I could not imagine a more important place in North America to concentrate conservation efforts," he said of the delta. "It's the crown jewel of conservation potential on the continent. Any large change in the ecosystem in the delta, though, has to come from the way the delta is managed. It won't come from small projects, really, like pulling up tamarisk, or planting willows. It must come from changing the way the river is managed."

Still, such restoration projects demonstrate that the delta has come a long way in only a few years of attention and care. Despite the dangers facing the delta, I find myself optimistic.

The four of us walk through grasses and brush and out onto the edge of the mudflats. It's a gooey mess. We sink up to our ankles in the salt-crusted muck, and then squeeze through a barbed-wire fence into another protected pond area. Cow tracks completely encircle the fence, but inside it is an undisturbed shallow pond surrounded by green grasses. María and Miriam begin using their long-handled nets to catch pupfish for their research. The fish scuttle about in very shallow waters on the edges, much easier to catch here than at Cerro Prieto.

Floating on the shallow pond, on the far side is an eared grebe. These are the same birds that died in such huge numbers in 1992 in the Salton Sea. They are a small waterbird with a pointed bill and lobed feet—they look like a cross between a duck and a loon. Because they are preeminently skilled at diving and catching fish, their legs are far back on their bodies, good for getting strong kicks and powerful propulsion. It means they are miserable at walking on land, and you almost never see them anywhere but on water or under it.

Of the various species of grebes, the eared grebe is one of the small ones. It occurs throughout the western half of the continent, and especially along the Pacific coast. It has a chunky little body with lovely chestnut flanks. Its back and neck and head are an obscure black. This one is in full breeding plumage and has spectacular markings on its face. The head has a high, black nubby crown of feathers. On each side of its head is a golden flare of feathers, really wonderful. They seem to flash when they catch the light right. They give the bird its name, "eared" grebe. At the center of each side of its face, and out of which the golden feathers seem to flow, are its eyes, which are as vital and living red as small pools of fresh-spilled blood.

*"Pato marijuano,"* Juan says, when I ask him what this bird is called here in the delta. Marijuana duck, named for its vivid bloodshot eyes. We both laugh.

As you know if you have ever tried to get close to a grebe on a lake, they are skittish and always flee, usually by dropping straight under water like a rock. But this one lets me walk up close to it. I tramp quietly and cautiously into the middle of the pond, up over my calves in muck and springwater. The grebe is nervous at first. It cannot dive, since the water is so shallow. It cannot walk either. And it does not fly, which surprises me.

Instead, it swims nervously near the reeds. I set up my camera and tripod in the middle of the water, arrange a very long lens, and begin to photograph. A thick cover of broken clouds obscures the sun. As the sun drops toward the horizon over the Sierra Cucapá, a rich yellow light honeys up the scene, filling it with rich, warm, sweet light.

You rarely see a wild creature so close up, and I cherish these happy moments. The bird does not leave, but actually begins to swim closer to me, as I simply stay in one location. Perhaps it does not fly, I think at that time, because it can't. Maybe it has selenium poisoning? I still don't know.

Soon it is almost close enough for me to touch. It swims within a couple feet of me, staring straight at me with those powerful red eyes. Nothing in nature is more arresting for me than being caught in the stare of another creature. The gaze of animals has a magical effect. It is not different from realizing that some other person is looking at us carefully. The intimacy of such moments is incomparable. But they always have an edge, too, at once mysterious, but full of meaning. Inevitably the look of the animal locates me in the world, in *its* world, not mine.

In the grebe's eyes I see myself being seen. I exist in the midst of other geographies and other lives, looking at me, looking at us. The moment inevitably disturbs the sense of human self-centeredness in the world. The fantasy that the world exists solely for us humans is irrevocably and suddenly pierced.

This red-eyed eared grebe does not inhabit the delta of fantasy. The bird is a living fact in the delta. This is what the Mexican delta is finally all about for those of us who know it and love it—acknowledging the reality of these other lives, animal and human, and their right to coexist with us.

The eared grebe seems not quite sure of what to make of me, what to do with me. It darts about me several times and then discovers a channel in the pond. With a rush of energy, it swims right past.

But I am still absorbed in its face. The flashing yellow feathers seem to stream out of the glowing red eye, like the tail from a comet. Or like the bright rays of a fierce red delta sun.

▲ *An eared grebe, stunning in its breeding plumage, swims amid shallows and reeds in El Doctor.*

285

# Appendix

## Endangered, Threatened, and Other Sensitive Species of the Lower Colorado River and the Delta of the Colorado River in Mexico

The following list of species is developed from the draft Conservation Plan for the Lower Colorado River Multi-Species Conservation Program (SAIC/Jones and Stokes, 2002). For a more complete picture of the condition of species in the area, species protected by the United States and Mexico that occur in the delta of the Colorado River and the Sea of Cortez have been added. It is important to note that the species listed as protected in Mexico is not exhaustive. Much study is currently under way to determine the status of many species in the Mexican delta, including the southwest willow flycatcher and the western yellow-billed cuckoo (possible nesting species in northwestern Mexico). The listing of a species in or by only one country indicates not only the need for more studies. It also indicates the historical lack of coordination between the two countries in environmental protection.

| Common and Scientific Name | Mexican Status[1] | U.S. Federal Status[2] | Arizona Status[3] | California Status[4] | Nevada Status[5] |
|---|---|---|---|---|---|
| **MAMMALS** | | | | | |
| Mexican long-tongued bat *Choeronycteris mexicana* | A | — | ASC | CSC | — |
| California leaf-nosed bat *Macrotus californicus* | — | — | ASC | CSC | — |
| Pallid bat *Antrozous pallidus* | — | — | — | CSC | — |
| Pale Townsend's big-eared bat *Corynorhinus townsendii pallescens* | — | — | ASC | CSC | — |
| Spotted bat *Euderma maculatum* | R | — | ASC | CSC | NP |
| Allen's big-eared bat *Indionycteris (=Plecotus) phyllotis* | — | — | — | — | — |
| Western (desert) red bat *Lasiurus blossevillii* | — | — | ASC | — | — |
| Western yellow bat *Lasiurus xanthinus* | — | — | — | — | — |
| Occult little brown bat *Myotis lucifugus occultus* | — | — | ASC | CSC | — |
| Cave myotis *Myotis velifer* | — | — | ASC | CSC | — |
| Greater western mastiff bat *Eumops perotis californicus* | — | — | ASC | CSC | — |
| Big free-tailed bat *Nyctinomops macrotis* | — | — | — | CSC | — |
| Desert pocket mouse *Chaetodipus penicillatus sobrinus* | — | — | — | — | — |
| Colorado River cotton rat *Sigmodon arizonae plenus* | — | — | — | CSC | — |
| Yuma hispid cotton rat *Sigmodon hispidus eremicus* | — | — | — | CSC | — |

| Common and Scientific Name | Mexican Status[1] | U.S. Federal Status[2] | Arizona Status[3] | California Status[4] | Nevada Status[5] |
|---|---|---|---|---|---|
| Ringtail *Bassariscus astutus* | — | — | — | FP | — |
| Vaquita porpoise *Phocoena sinus* | Ex | E | — | — | |
| Jaguar *Panthera onca* | Ex | E | | | |
| Blue whale *Balaenopkra musculus* | SP | — | — | — | — |
| Fin whale *Balaenoptera physalus* | SP | — | — | — | — |
| California gray whale *Eschrichtius robustus* | SP | — | — | — | — |
| **BIRDS** | | | | | |
| Clark's grebe *Aechmophorus clarkii* | — | — | — | — | — |
| California brown pelican *Pelecanus occidentalis* | — | E | — | CE/FP | — |
| Western least bittern *Ixobrychus exilis hesperis* | — | — | — | CSC | — |
| Great blue heron *Ardea herodias* | R | — | — | — | — |
| Reddish Egret *Egretta rufescens* | A | — | — | — | — |
| Brant *Branta bernicla* | SP | — | — | — | — |
| Fulvous whistling-duck *Dendrocygna bicolor* | — | — | — | CSC | — |
| Golden eagle *Aquila chrysaetos* | — | — | — | CSC/FP | — |
| Swainson's hawk *Buteo swainsoni* | — | — | — | CT | — |
| Bald eagle *Haliaeetus leucocephalus* | Ex | T | — | CE/FP | NE |
| Harris' hawk *Parabuteo unicinctus* | A | — | — | CSC | — |
| American peregrine falcon *Falco peregrinus anatum* | A | — | — | CE/FP | NE |
| California black rail *Laterallus jamaicensis coturniculus* | — | — | — | CT/FP | — |
| Yuma clapper rail *Rallus longirostris yumanensis* | Ex | E | — | CT/FP | — |
| Greater sandhill crane *Grus canadensis tabida* | — | — | — | CT/FP | — |
| Mountain plover *Charadrius montanus* | A | PT | — | — | — |
| Heermann's gull *Larus heermannii* | A | — | — | — | — |

| Common and Scientific Name | Mexican Status[1] | U.S. Federal Status[2] | Arizona Status[3] | California Status[4] | Nevada Status[5] |
|---|---|---|---|---|---|
| Elegant tern *Sterna elegans* | A | — | — | — | — |
| Western yellow-billed cuckoo *Coccyzus americanus occidentalis* | — | — | — | CE | — |
| Burrowing owl *Athene cunicularia* | — | — | — | CSC | — |
| Elf owl *Micrathene whitneyi* | — | — | — | CE | — |
| Gilded flicker *Colaptes chrysoides* | — | — | — | CE | — |
| Blue footed booby *Sula nebouxii* | A | — | — | — | — |
| Gila woodpecker *Melanerpes uropygialis* | — | — | — | CE | — |
| Southwestern willow flycatcher *Empidonax trailii extimus* | — | E | — | CE | — |
| Vermilion flycatcher *Pyrocephalus rubinus* | — | — | — | CSC | — |
| Arizona Bell's vireo *Vireo bellii arizonae* | — | — | — | CE | — |
| Northern mockingbird *Mimus polyglottos* | SP | — | — | — | — |
| Crissal thrasher *Toxostoma crissale* | — | — | — | CSC | — |
| Sonoran yellow warbler *Dendroica petechia sonorana* | — | — | — | CSC | — |
| Summer tanager *Piranga rubra* | — | — | — | CSC | — |
| Large-billed savannah sparrow *Passerculus sandwichensis rostratus* | R | — | — | CSC | — |
| Northern cardinal *Cardinalis cardinalis* | — | — | — | CSC | — |
| House finch *Carodacus mexicanus* | SP | — | — | — | — |
| **REPTILES** | | | | | |
| Desert tortoise (Mojave population) *Gopherus agassizii* | A | T | ASC | CT | NT |
| Flat-tailed horned lizard *Phrynosoma mcalli* | A | PT | ASC | CSC | — |
| **AMPHIBIANS** | | | | | |
| Colorado river toad *Bufo alvarius* | — | — | — | CSC | — |
| Relict leopard frog *Rana onca* | — | — | ASC | — | NSP |
| Lowland leopard frog *Rana yavapaiensis* | R | — | ASC | CSC | — |

| Common and Scientific Name | Mexican Status[1] | U.S. Federal Status[2] | Arizona Status[3] | California Status[4] | Nevada Status[5] |
|---|---|---|---|---|---|
| **FISH** | | | | | |
| Colorado Pikeminnow<br>*Ptychocheilus lucius* | Ex | E | — | CE | — |
| Bonytail chub<br>*Gila elegans* | Ex | E | ASC | CE | NE |
| Humpback chub<br>*Gila cypha* | — | E | ASC | — | NE |
| Flannelmouth sucker<br>*Catostomus latipinnis* | — | — | ASC | — | — |
| Razorback sucker<br>*Xyrauchen texanus* | Ex | E | ASC | CE/FP | NE |
| Desert pupfish<br>*Cyprinodon macularius* | Ex | E | ASC | CE | — |
| Totoaba<br>*Cynoscion macdonaldii* | Ex | E | — | — | — |
| **INVERTEBRATES** | | | | | |
| MacNeil's sooty winged skipper<br>*Pholisora gracielae* | — | C | — | — | — |
| **PLANTS** | | | | | |
| Las Vegas bearpoppy<br>*Arctomecon california* | — | — | AP | — | NEP |
| Sticky buckwheat<br>*Eriogonum viscidulum* | — | — | — | — | NEP |
| Threecorner milkvetch<br>*Astragalus geyeri* var. *triquetrus* | — | — | — | — | NEP |

[1]**Mexican Status** (four legal categories afforded by Normas Officiales Mexicanas (NOM—ECOL—059—1994)
  Ex=En peligro de extinción (endangered)
  A=Amenazada (threatened)
  R=Rara (Rare)
  SP=Sujeta a protección especial (Species of Special Concern)

[2]**U.S. Federal Status**
  *E=Listed as endangered under the federal Endangered Species Act (ESA)*
  T=Listed as threatened under ESA
  PT=Proposed for listing as threatened under ESA
  PE=Proposed for listing as endangered under ESA

[3]**Arizona Status**
  AP=Arizona Protected Plant
  ASC=Arizona Wildlife of Special Concern

[4]**California Status**
  CE=Listed as endangered under the California Endangered Species Act (CESA)
  CT=Listed as threatened under CESA
  FP=Fully protected under the California Fish and Game Code
  CSC=California species of special concern.

[5]**Nevada Status**
  NT=Nevada threatened
  NEP=Nevada Critically Endangered Plant
  NP=Nevada Protected

# Bibliography and Reference List

Abarca, F. J., M. F. Ingraldi, and A. Vareli-Romero. 1993. "Observaciones del Cachorrito del Desierto (*Cyprindodon macularius*), Palmoteador de Yuma (*Rallus longirostris yumanenesis*) y Comunidades de Aves Playeras en la Ciénega de Santa Clara, Sonora Mexico."In *Nongame and Endangered Wildlife Program Technical Report*. Phoenix: Arizona Fish and Game Department.

Aboites Aguilar, Luis. 1998. *El Agua de la Nación: Una Historica Política de México (1888–1946)*. México, D.F.: Centro de Investigaciones y Estudios Superiores en Antropología Social.

Agreement. 1983. "Agreement between the United States of America and the United Mexican States on Cooperation for the Protection and Improvement of the Environment in the Border Area." 14 August, La Paz, México.

Aguirre Bernal, Celso. 1966. *Compendio Histórico-Biográfico de Mexicali 1537–1966*. Mexicali, México: n.p.

Aguirre, Amado. 1977. *Documentos para la Historia de Baja California*. México, D.F.: Instituto de Investigaciones Históricas, Universidad Nacional Autónoma de México.

Alarcón. *The relation of the navigation and discovery which Captaine Fernando Alarchon made by order of the right honourable Lord Don Antonio de Mendoza Vizeroy of New Spaine, dated in Colima, an haven of New Spaine*. Reprinted from Hakluyt's "Principal Navigations" in Hakluyt Soc. Publs., Extra Series, Vol. 9, Glasgow, 1904, pp. 279–318.

Albrecht, Virginia S., John J. Rhodes III, and Andrew J. Turner. 2001. "Memorandum of Amici Central Arizona Water Conservation District, Coachella Valley Water District, Imperial Irrigation District, Metropolitan Water District of Southern California, San Diego County Water Authority and Arizona Power Authority in Support of Defendants' Opposition to Plaintiffs' Motion for Summary Judgment and in Support of Defendants' Cross Motion for Summary Judgment." United States District Court for the District of Columbia, no. 1:00CV01544 (JR). Filed August 3.

Alemán-Romeros, R. L., y S. A. Ochoa-Sanchez. 1994. *Diagnóstico Socioeconómico de las Comunidades Pesqueras del Alto Golfo de California*. Informe Global. Comité Técnico para la Preservación de la Vaquita Marina y la Totoaba. Ensenada: INP-CRIP.

Alvarez, Juan, and Víctor M. Castillo. 1986. *Ecología y frontera: ecology and the borderlands*. N.p.: Universidad Autónoma de Baja California.

Alvarez-Borrego, Saúl. 1983. "The Gulf of California." In *Estuaries and Enclosed Seas*, Ecosystems of the World, vol. 26, ed. Bostwick H. Ketchum, 427–49.

Amsterdam: Elsvervier Scientific Publishing Company.

Alvarez-Borrego, Saúl. 1999a. "Physical and Biological Linkages between the Upper and Lower Colorado Delta." Paper presented presented at the Third International Congress on Ecosystem Health, PLACE OF MEETING, 15–20 August.

Alvarez-Borrego, Saúl. 1999b. "Water Quality in the Colorado River and Upper Gulf of California: An Environmental Binational Issue." Paper presented at Future of the Colorado River Basin Conference, 18 November, Mexicali, Mexico.

Alvarez-Borrego, Saúl, y L. A. Galindo-Bect. 1974. "Hidrología del Alto Golfo de California I. Condiciones del Otoño." *Ciencias Marinas* 1: 46–64.

Alvarez-Borrego, S., B. P. Flores-Baez, y L. A. Galindo-Bect. 1975. "Hidrología del Alto Golfo de California II. Condiciones durante Invierno, Primavera, y Verano." *Ciencias Marinas* 2: 21–36.

Alvarez-Borrego, Saúl, Daniel W. Anderson, Guillermo Ferández de la Garza, John Letey, Mark R. Matsumoto, Gerald T. Orlob, and Juan-Vicente Palerm. 1999. *Alternative Futures for the Salton Sea*. The UC MEXUS Border Water Project, Issue Paper no. 1. PLACE: University of California Institute for Mexico and the United States and the University of California Water Resources Center.

Alvarez de Williams, Anita. 1974. *The Cocopah People*. Phoenix: Indian Tribal Series.

———. 1975. *Travelers among the Cucapá*. Los Angeles: Dawson's Book Shop.

———. 1983. "Cocopa." In *Handbook of North American Indians*, vol. 10, ed. A. Ortiz, 99–110. Washington, D.C.: Smithsonian Institution.

———. 1989. "Parte 4: Historia. Los Indios Cucapá del Delta del Río Colorado, Historia de la Región de Puerto Peñasco." *Noticias del CEDO* 2: 16.

———. 1997. "People and the River." In *Dry Borders: A Double Issue*, special issue of *Journal of the Southwest* 39 (autumn–winter):331–51.

American Rivers. *1998 Most Endangered Rivers Report*. www.americanrivers.org.

Anderson, Daniel W., Edward Palacios, Eric Mellink, and Carlos Valdes-Casillas. 1999. "Migratory Bird Conservation in the Colorado River Delta Region." Paper presented at the Third International Congress on Ecosystem Health, Sacramento, California, 15–20 August.

Anguiano Téllez, María Eugenia. 1995. *Agricultura y migración en el Valle de Mexicali*. Tijuana, México: El Colegio de la Frontera Norte.

Angulo, César. 2000. "Atentan contra curvina." *La Crónica* (Mexicali, México), 7 Abril, 1A.

Aragón-Noriega, E. A., and L. A. Calderón Aguilera. 2000. "Does Damming of the Colorado River Affect the Nursery Area of Blue Shrimp Litopenaeus stylirostris (Decapoda: Penaeidae) in the Upper Gulf of California." *Revista de Biología Tropical* 49: 867–71.

Associated Press. 2001. "River Conservation Supported in Poll." *ENN.* 28 June.

Aubert, H., and M. Vásquez León. 1993. "The Ethnography of Fishing." In *Maritime Community and Biosphere Reserve: Crisis and Response in the Upper Gulf of California,* Bureau of Applied Research in Anthropology, paper no. 2, ed. T.R. McGuire and J. B. Greenberg, 49–75. Tucson: University of Arizona Press.

Audubon, John Woodhouse. [1906] 1969 *Audubon's Western Journal: 1849–1850.* Reprint, Glorieta, New Mexico: Rio Grande Press.

Austin, Mary. 1927. "The Colorado River Controversy." *The Nation* 125: 512–15.

Baars, Donald L. 1972. *The Colorado Plateau: A Geologic History.* Albuquerque: University of New Mexico Press.

Babbitt, Bruce and Julia Carrabias. 1997. "Letter of Intent between the Department of the Interior of the United States and the Secretariat of Environment, Natural Resources, and Fisheries (SEMARNAP) of the United Mexican States for Joint Work in Natural Protected Areas on the U.S.–Mexico Border. May 5, Washington, D.C. Unpublished manuscript.

———. 2000. "Joint Declaration between the Department of the Interior of the United States of America and the Secretariat of Environment, Natural Resources, and Fisteries (SEMARNAP) of the United Mexican States to Enhance Cooperation on the Delta of the Colorado River." 18 May, Washington, D.C. Unpublished manuscript.

Bachelard, Gaston. [1964] 1994 *The Poetics of Space.* Trans. Maria Jolas. Reprint, Boston: Beacon Press.

Bancroft, G. 1922. "Some Winter Birds of the Colorado Delta." *Condor* 24: 98.

Bancroft, H. H. 1883. *History of the Pacific States of North America, Vol. 10: North Mexican States, Vol. 1 (1531–1800).* San Francisco.

Banks, R. C., and R. E. Tomlinson. 1974. "Taxonomic Status of Certain Clapper Rails of Southwestern United States and Northwestern Mexico." *Wilson Bulletin* 86: 325–35.

Barrera-Guevara, J.C. 1990. "The Conservation of *Totoaba macdonaldi* (Gilbert), (Pisces: Sciaenidae), in the Gulf of California, Mexico." *Journal of Fish Biology* 37, Supplement A: 201–02.

Barrera-Guevara, J.C., y J. Campoy. 1992. "Ecología y Conservación del Alto Golfo de California." En *Ecología, Recursos Naturales y Medio Ambiente en Sonora,* comp. J. L. Moreno. Sonora, México: Coed. Colegio de Sonora y Gobierno del Estado.

Barrows, David P. 1900. "The Colorado Desert." *National Geographic* 11(Sept.): 336–51.

Bate, Jonathan. 2000. *The Song of the Earth.* London: Picador.

Berdegué, A. J. "La Pesquería de la Totoaba (*Cynoscion macdonaldi*) en San Felipe, Baja California. *Revista de la Sociedad Mexicana de Historia Natural* 34: 293–300.

Berwyn, Bob. 2001. "Greens Want Colorado River Ecosystem Restored." March 8, www.enn.com/news/enn-stories/2001/03082001/river_42403.asp?site=e.

"Binational Declaration—Declaración Binacional: The Colorado River Delta—El Delta del Río Colorado." 2001. Symposium on the Colorado River. Mexicali, Mexico, 12 September. Unpublished manuscript.

Blackwater, E. 1934. "Origin of the Colorado River." *Bulletin of the Geological Society of America* 36: 551–66.

Blake, W. P. 1909. "Lake Cahuilla: The Ancient Lake of the Colorado." *Out West* 30: 74.

Blanco, Jacobo. 1873. "Informe sobre la Exploración del Río Colorado." México: Ministerio de Fomento.

Bolton, Herbert Eugene, ed. 1916. *Spanish Exploration in the Southwest 1542–1706.* Original Narratives in Early American History. New York: Charles Scribner's Sons.

———. 1919. "Father Escobar's Relation of the Oñate Expedition to California." *Catholic Historical Review* 5: 15–41.

———. 1930. *Anza's California Expeditions.* 5 vols. Berkeley: University of California Press.

———. [1936] 1984. *Rim of Christendom: A Biography of Eusebio Francisco Kino, Pacific Coast Pioneer.* Reprint, Tucson: University of Arizona Press.

———. [1949]. *Coronado: Knight of Pueblos and Plains.* Albuquerque: University of New Mexico Press.

Bonillas, Y. S., y F. Urbina. 1912–1913. "Informe acerca de los Recursos Naturales de la Parte Norte de la Baja California, especialmente del Delta del Río Colorado." *Paragones del Instituto Geológico de México* 4: 161–235.

Boslough, John. 1981. "Rationing a River: Everyone's Been Promised a Piece of the Colorado, but There's Not Enough to Go Around." *Science* 81 (2): 26–36.

Bowden, Charles. 1977. *Killing the Hidden Waters.* Austin: University of Texas Press.

Boyle, Robert H. 1996. "Life—or death—for the Salton Sea? Large Polluted California Lake Has Increasing Salinity and Pollution." *Smithsonian* 27 (June): 86–96.

Briggs, Mark K. 1996. *Riparian Ecosystem Recovery in Arid Lands: Strategies and References.* Tucson: University of Arizona Press.

Briggs, Mark K., and Steven Cornelius. 1998. "Opportunities for Ecological Improvement Along the Lower Colorado River and Delta." *Wetlands* 18): 513–29.

Brody, Jane. 1997. "Hope Rising for Selenium." *New York Times,* February 19, 1997.

Brown, B. T. 1988. "Breeding Ecology of a Willow Flycatcher Population in Grand Canyon, Arizona." *Western Birds* 19: 25–33.

Brown, D. E. 1982. "Biotoic Communities of the American Southwest—United States and Mexico." *Desert Plants* 4: 288 ff.

Brownell, R. L. 1986. "Distribution of the Vaquita

*Phocoena sinus,* in Mexican Waters." *Marine Mammals Sciences* 2: 299–305.

Bryne, J. U., and K. O. Emery. 1960. "Sediments of the Gulf of California." *Geological Society of America Bulletin* 71: 983–1010.

Burnett, E., E. Kandl, and F. Croxen. [1993] 1997. *Cienega de Santa Clara: Geologic and Hydrologic Comments.* Revised, Yuma, Arizona: U.S. Bureau of Reclamation, Yuma Projects Office.

Burton-Christie, Douglas. 1999. "Faint Music: Language, Landscape and the Sacred." *Ascent* 24 (fall): 44–51.

———. 2000. "Words Beneath the Water: Logos, Cosmos, and the Spirit of Place." In *Christianity and Ecology: Seeking the Well-Being of the Earth and Humans,* ed. Dieter T. Hessel and Rosemary Radford Ruether, 317–36. Cambridge, Massachusetts: Harvard University Press.

Cabeza de Vaca, Alvar Nuñez. [1961] 1983. *Adventures in the Unknown Interior of America.* Ed. and trans. Cyclone Covey. New York: Collier Books. Reprint, Albuquerque: University of New Mexico Press.

Calderón Salazar, Jorge A. 1990. *Reforma Agraria y Colectivización Ejidal en México: La experiencia cardenista.* Culiacán, México: Universidad Autónoma de Sinoloa.

Carrier, Jim. 1991. "The Colorado: A River Drained Dry." *National Geographic* 182 (June): 4–35.

Carriquiry, J. D., and A. Sánchez. 1999. "Sedimentation in the Colorado River Delta and Upper Gulf of California after Nearly a Century of Discharge Loss." *Marine Geology* 158: 125–45.

Chapman, Charles E. 1921. *A History of California: The Spanish Period.* New York: Macmillan.

Chase, J. Smeaton. 1919. *California Desert Trails.* Boston: Houghton Mifflin.

Chittenden, Newton H. 1901. "Among the Cocopahs." *Land of Sunshine* 41: 196–204.

Chomsky, Noam, and David Barsamian. 2001. *Propoganda and the Public Mind: Conversations with Noam Chomsky.* Boston: South End Press.

Christie, Joyce. 1996. "Desert Wetlands: Mexico Delta Gets New Life When Water Returns." *Yuma Valley Sun,* 4 April, 1C, 11C.

Cisneros, M. A., G. Montemayor, and M. Román. 1995. "Life History and Conservation of *Totoaba macdonaldi.*" *Conservation Biology* 9: 806–14.

Clark, Jo, et al. 2001. "Immediate Options for Augmenting Water Flows to the Colorado River Delta in Mexico." Presented to the David and Lucile Packard Foundation, May. Stet. Add: Unpublished manuscript.

Clifford, Frank. 1997a. "Efforts Focus on the Dried-Up Colorado Delta." *Albuquerque Journal,* 30 March.

———. 1997b. "Plotting a Revival in a Delta Gone to Dust." *Los Angeles Times,* 24 March, A-1.

Clifford, James. 1997. *Routes: Travel and Transition in the Late Twentieth Century.* Cambridge, Massachusetts: Harvard University Press.

Cohen, Michael, Jason I. Morrison, and R. Glennon. 1999a. "Conservation Value and Water Management Issues of the Wetland and Riparian Habitats in the Colorado River Delta in Mexico." Paper presented at the Third International Congress on Ecosystem Health, Sacramento, California, 15–20 August.

———. 1999b. *Haven or Hazard: The Ecology and Future of the Salton Sea.* Oakland, California: The Pacific Institute.

Cohen, Michael, C. Henges-Jeck, and G. Castillo-Moreno. 2001. "A Preliminary Water Balance for the Colorado River Delta, 1992–1998." *Journal of Arid Environments* 49: 35–48.

Colleta, Bernard, and Luc Ortlieb. 1984. "Deformities of Middle and Late Pleistocene Deltaic Deposits at the Mouth of the Río Colorado, Northwestern Gulf of California." Paper presented at Neotectonics and Sea Level Variations in the Gulf of California Area, a Symposium. Hermosillo, Mexico, 21–23 April.

Comité Técnico para la Preservación de la Vaquita y la Totoaba (CTPVT). 1993. "Propuesta para la Declaración de Reserva de la Biosfera Alto Golfo de California y Delta del Río Colorado." Documento inédito.

Comisión Nacional del Agua (CONAGUA). 1994. *Hidrología y Administración del Río Colorado.* México, D.F.: Editorial Trillas.

Consag, Ferdnando. 1759. "Account of the Voyage of Father Fernando Consag, Missionary of California, Performed for Surveying in the Eastern Coast of California to Its Extremity, the River Colorado, by Order of Father Christoval de Escobar and Llamas, Provincial of New Spain, in the Year 1746." In *A Natural and Civil History of California . . . Translated from the Original Spanish of Miguel Venegas. . . .* 2 vols. London.: Vol. 2, appendix 3, 308–53.

Cooper, J. G. 1869. "The Naturalist in California. The Colorado Valley in Winter." *American Naturalist* 3: 470–81.

Coronado, Irasema. 1999. "Conflicto por el Agua en la Región Fronteriza." *Borderlines* 7 (Julio): 1–4.

Cory, H. T. 1915. *The Imperial Valley and Salton Sink.* San Francisco.

Coues, Elliott, ed. 1900. *On the Trail of a Spanish Pioneer: Garcés Diary 1775–1776.* 2 vols. New York: Harper.

"Courses of Action Identified at the Symposium on the Delta of the Colorado River." 2001. Symposium on the Colorado River. Mexicali, México, 12 Septiembre. Unpublished manuscript.

Cox, Gail Diane. 1993. "Status Quo Is Threatened on Nation's Most Litigated River." *National Law Journal* 16 (13 September).

Coyro, Ernesto Enriquez. 1976. *El Tratado entre México y Los Estados Unidos de América sobre Ríos Internacionales: Una Lucha Nacional de Noventa Años.* 2 vols. México, D.F.: Universidad Nacional Autónoma de México.

———. 1984. *Los Estados Unidos de América ante Nuestro Problema Agrario.* México, D.F.: Universidad Nacional Autónoma de México.

*La Crónica* (Mexicali, México). 2000. "Piden al presidente Zedillo conserve Delta del Colorado." 6 Junio, 1A.

Crosby, Alfred W. 1986. *Ecological Imperialism: The Biological Expansion of Europe, 900–1900*. Cambridge, England: Cambridge University Press.

Cruz, Francisco Santiago. 1969. *Baja California, Biografía de Una Península*. México, D.F.: Editorial Jus.

Cudney Bueno, R., y P. J. Turk Boyer. 1998. *Pescando entre Mares del Alto Golfo de California. Una Guía sobre la Pesca Artesanal, Su Gente y Sus Propuestas de Manejo*. Puerto Peñasco: Centrao Intercultural de Estudios de Desiertos y Oceános, (CEDO).

Culp, Peter W. 2000. *Restoring the Colorado Delta with the Limits of the Law of the River: The Case for Voluntary Water Transfers.*, City: Udall Center for Studies in Public Policy, University of Arizona.

D'Agrosa, Caterina, and Omar Vidal. 2000. "Vaquita or Something Similar." *Conservation Biology* 14: 1110–19.

Darrah, William Culp. 1951. *Powell of the Colorado*. Princeton: Princeton University Press.

Davila, Ismael. 2000. "Denuncian indígenas represión." *El Mexicano* (Mexicali, México), 4 Abril, 1A, 2A.

Davis, Owen K., Alan H. Cutler, Keith H. Meldahl, Joseph F. Schreiber, Jr., Brian E. Lock, Lester J. Williams, Nicholas Lancaster, Christopher A. Shaw, and Stephen M. Sinitiere.1990. "Quaternary Geology of Bahai Adair and the Gran Desierto Region." In *Deserts: Past and Future Evolution*, ed. Owen K. Davis, 132. IGCP 252.

Day, Allen. 1906. "The Inundation of the Salton Sea by the Colorado River and How It Was Caused." *Scientific American* 94, 310–12.

deBuys, William, and Joan Myers. 1999. *Salt Dreams: Land and Water in Low-Down California*. Albuquerque: University of New Mexico Press.

Defenders of Wildlife, et al. 2000. *Defenders of Wildlife et al. v. Norton et al.* United States District Court for the District of Columbia, no. 1:00CV01544. Filed June 8.

Delgado, Jaime. 1998. "Entregan indígenas 'rosario' de necesdidades a Gobernador." *La Crónica* (Mexicali, México), 16 Diciembre, 4A.

Dellapenna, Joseph W. 2001. "The Customary International Law of Transboundary Fresh Waters." *Journal of International Water*, 1: 1–43. May.

Dellenbaugh, F. S. 1902. *The Romance of the Colorado River*. New York.

Derby, George H. 1852. *Reconnoissance [sic] of the Gulf of California and the Colorado River*. 32nd Cong., 1st sess., Senate Executive Document no. 81. Washington, D.C.: Government Printing Office.

DeVoto, Bernard. 1947. "The West against Itself." *Harper's*, 194 (January): 2.

Díaz, Melchior. 1896. In the first part of *Relacion de la Jornada de Cibola conpuesta por Pedro de Casteñada de Nazera donde se trata de todos aquellos poblados y ritos y costumbres, la qual fue el Año de 1540*. 14th Annual Rept. Bur. of Ethnol. 1892–93, 414–69 in Spanish, trans. G. P. Winship in 470–546, Dillin, John. 1989. "Pollution Seeps from Mexico to U.S." *Christian Science Monitor*, 28 December, 6.

Donahue, John M., and Barbara Rose Johnston. 1998.

*Water, Culture, and Power: Local Stuggles in a Global Context*. Washington, D.C.: Island Press.

Dowd, Munson J. 1960. *Historic Salton Sea*. N.p.: Imperial Irrigation District.

Duncan, David James. 2001. *My Story as Told by Water: Confessions, Druidic Rants, Reflections, Bird-Watchings, Fish-Stalkings, Visions, Songs and Prayers Refracting Light, from Living Rivers, in the Age of Industrial Dark*. San Francisco: Sierra Club Books.

Dutton, Clarence. 1882. *Tertiary History of the Grand Canyon District*. U. S. Geological Survey Monograph 2. Washington, D.C.

Endangered Species Act. 1973. www.fws.gov/~r9endspp/esa.htm.

Eddleman, W. R. 1989. *Biology of the Yuma Clapper Rail in the Southwestern United States and Mexico. Final Report*. Inter-Agency Agreement no. 4-A. Yuma, Arizona: Bureau of Reclamation, Yuma Projects Office.

Eddleman, W. R., R. E. Flores, and M. L. Lee. 1994. "Black Rail, *Laterallus jamaicensis*." In *The Birds of North America, Life Histories for the 21st Millenium*, no. 123, ed. A. Poole and F. Gill. Washington, D.C.: American Ornithologists Union.

Eddlemaan, W. R., and J. C. Courtney. 1998. "Clapper Rail, *Rallus longirostris*." In *The Birds of North America, Life Histories for the 21st Millenium*, no. 340, ed. A. Poole and F. Gill. Washington, D.C.: American Ornithologists Union.

Environmental Protection Agency. 1996. "U.S.–Mexico Border XXI Program, Framework Document. Environmental Protection Agency, Washington, D.C.

Egan, Timothy. 2001. "Water: The New Liquid Gold." *New York Times*, August 12, page.

Evernden, Neil. 1992. *The Social Creation of Nature*. Baltimore: Johns Hopkins University Press.

Ezcurra, Exequiel, Luis Bourillon, Antonio Cantú, María Elena Martínez, and Alejandro Robles. "Ecological Conservation in the Sea of Cortés." In *Island Biogeography in the Sea of Cortés*, ed. M. L. Cody, T. Case, and E. Ezcurra. Oxford University Press, in press.

Ezcurra, Exequiel, R. S. Felger, A. D. Russell, and M. Equihua. 1988. "Freshwater Islands in a Desert Sea: The Hydrology, Flora, and Phytogeography of the Gran Desierto Oases of Northwestern Mexico." *Desert Plants* 9: 35–63.

Felger, R. S. "Vegetation and Flora of the Gran Desierto, Sonora, Mexico." *Desert Plants* 2: 87–114.

———. 1992. "Syposis of the Vascular Plants of Northwestern Sonora, Mexico. *Ecologica* 2: 11–44.

Findley, Rowe. 1973. "The Bittersweet Waters of the Lower Colorado River." *National Geographic* Magazine 164 (October): 540–69.

Flanagan, C. A., and J. R. Hendrickson. 1976. "Observation on the Reproductive Biology of the Totoaba, *Cynocion macdonaldi*, in the Gulf of California, Mexico." *Fisheries Bulletin* 74: 531–44.

Foltz, Bruce V. 1995. *Inhabiting the Earth: Heidegger, Environmental Ethics, and the Metaphysics of Nature*. Atlantic Heights, New Jersey: Humanities Press.

Fradkin, Philip L., Jr. 1996. *A River No More: The Colorado River and the West.* Berkeley: University of California Press; originally publ. 1968.

Freeman, Lewis R. 1930. *Down the Grand Canyon.* New York: Dodd, Mead.

Friederici, Peter. 1998. "The Colorado Loses Out: Fish and Wildlife Suffer as Dams Divert Water to Crops, Lawns and Golf Courses in the Desert." *Defenders,* 73 (spring): 10–18, 31–33.

Funcke, E. W. 1915. "The Sheep Hunting of Lower California." *Outdoor Life* 32: 221–28.

Galindo-Bect, Manuel S., et al. "El Río Colorado: Componente Vital en el Area de Crianza del Alto Golfo de California." Documento inédito.

Galindo-Bect, Manuel S., Edward P. Glenn, et al. 2000. "Penaeid Shrimp Landings in the Upper Gulf of California in Relation to Colorado River Freshwater Discharge." *Fisheries Bulletin* 98: 222–25.

Ganster, Paul. 1996. "Environmental Issues of the California-Baja California Border Region." Border Environmental Research Reports, no. 1. San Diego: Southwest Center for Environmental Research and Policy, San Diego State University, June.

Garces, Fr. Francisco. 1955. *A Record of Travels in Arizona and California.* Trans. John Galvin. Place if known: John Howell Books.

García Moreno, Víctor Carlos, comp. 1982. *Análisis de algunos problemas fronterizos y bilaterales entre México y Estados Unidos.* México, D.F.: Universidad Nacional Autónoma de México.

García-Hernandez, Jaqueline, Edward P. Glenn, Janick Artiola, and Don J. Baumgartner. 2000a. "Bioaccumulation of Selenium (Se) in the Cienega de Santa Clara Wetland, Sonora, Mexico." *Ecotoxicology and Environmental Safety* 46: 298–304.

García-Hernandez, Jaqueline, Osvel Hinojosa-Huerta, EdwardP. Glenn, Vanda Gerhart, and Yamilett Carrillo-Guerrero. 2000b. "Southwestern Willow Flycatcher Survey in Cocopah Territory, Yuma, Arizona." Final report prepared for the Cocopah Indian Tribe. Tucson: Unpublished manuscript. .

García-Hernandez, Jaqueline, Kirke A. King, Anthony L. Velasco, Evgueni Shumlin, Miguel A. Mora, and Edward P. Glenn. 2001a. "Selenium, Selected Inorganic Elements, and Organochlorine Pesticides in Bottom Material and Biota from the Colorado River Delta." *Journal of Arid Environments* 49: 65–90.

García-Hernandez, Jaqueline, Osvel Hinojosa-Huerta, Vanda Gerhart, Yamilett Carrillo-Guerrero, and Edward P. Glenn. 2001b. "Willow Flycatcher (*Empidonax traillii*) Surveys in the Colorado River Delta: Implications for Management." *Journal of Arid Environments* 49: 161–70.

Gates, Marilyn. 1988. "Codifying Marginality: The Evolution of Mexican Agricultural Policy and Its Impact on the Peasantry." *Journal of Latin American Studies* 20 (November): 277–311.

Gehlbach, Frederick R. [1981] 1993. *Mountain Islands and Desert Seas: A Natural History of the U.S.–Mexican Borderlands.* Reprint, College Station, Texas: Texas A and M Press.

Getches, David H. 1997. "Colorado River Governance: Sharing Federal Authority as an Incentive to Create a New Institution." *University of Colorado Law Revi* 68 : 573–658.

Gleick, Peter H. 1993. "Water and Conflict: Fresh Water Resources and International Security." *International Security* 18: 79–112.

———. 2001. "Making Every Drop Count." *Scientific American* 284 (February): 40–45.

Glenn, Edward P., Richard S. Felger, Alberto Búrquez, and Dale S. Turner. 1992. "Ciénega de Santa Clara: Endangered Wetland in the Colorado River Delta, Sonora Mexico." *Natural Resources Journal* 32 (fall): 817–24.

Glenn, Edward P., Christopher Lee, Richard Felger, and Scott Zengel. 1996. "Water Management Impacts on the Wetlands of the Colorado River Delta, Mexico." *Conservation Biology* 10 (August): 1175–86.

Glenn, Edward P., Chelsea Congdon, and Jaqueline Garcia. 1997. "New Value for Old Water." *The World and I* (April): 204–11.

Glenn, Edward P., and Carlos Valdes. 1998a. "Importance of United States Water Flows to the Colorado River Delta and the Northern Gulf of California, Mexico." Unpublished manuscript for Defenders of Wildlife and Southwest Center for Biological Diversity.

Glenn, Edward P., Rene Tanner, Shelby Mendez, Tamra Kehret, David Moore, Jaqueline García, and Carlos Valdes. 1998b. "Growth Rates, Salt Tolerance and Water Use Characteristics of Native and Invasive Riparian Plants from the Delta of the Colorado River, Mexico." *Journal of Arid Environments* 40: 281–94.

Glenn, Edward P., Michael Cohen, Jason I. Morrison, Carlos Valdes-Casillas and Kevin Fitzsimmons. 1999a. "Science and Policy Dilemmas in the Management of Agricultural Waste Waters: The Case of the Salton Sea, CA, USA." *Environmental Science and Policy* 2: 413–23.

Glenn, Edward P., R. Tanner, C. Congdon and D. Luecke. 1999b. "Status of Wetlands Supported by Drainage Water in the Colorado River Delta, Mexico." *HortScience* 34: 16–21.

Glenn, E. P., F. Zamora-Arroyo, P. Nagler, M. Briggs, W. Shaw, and K. Flessa. 2001. "Ecology and Conservation Biology of the Colorado River Delta, Mexico." *Journal of Arid Environments* 49: 5–16.

Glennon, Robert Jerome, and Peter W. Culp. 2001. "The Last Green Lagoon: How and Why the Bush Administration Should Save the Colorado River Delta." Unpublished manuscript.

Gomez-Pompa, Arturo. 1993. "Caught in Controversy: The Tuna, the Dolphin, and the 'Little Cow'." *UC Mexus News,* no. 31 (fall): 1–2.

Godínez-Plascencia, J. A., C. I. Vázquez-León, E. Martinéz, S. Romo-Zuñiga, y E. P. Vargas. 1994. "Evaluación Socioeconómica del Sector Pesquero del Alto Golfo de California." Comité Técnico para la Perservación de la Vaquita Marina. Tijuana, México.

Goodfriend, Glenn A., and Karl W. Flessa. 1997. "Radiocarbon Reservoir Ages in the Gulf of California: Roles of Upwelling and Flow from the Colorado River. *Radiocarbon* 39: 139–48.

Graf, William L. 1985. *The Colorado River: Instability and Basin Management.* Washington, D.C.: Association of American Geographers.

Graham, Frank Jr. 1998. "Midnight at the Oasis." *Audubon* 100 (May): 82–89.

Grinnell, Joseph. 1908. "Birds of a Voyage on Salton Sea." *The Condor* 10: 185–91.

———. 1914. "An Account of the Mammals and Birds of the Lower Colorado Valley, with a Special Reference to the Distributional Problems Presented." *University of California Publications in Zoology* 12: 51–264.

Grossfeld, Stan. 1997. "A River Runs Dry; A People Wither: Their Water Taken, Mexico's Cocopa Cling to Arid Homeland." *Boston Globe*, 21 September, A-1.

Grunsky, C. E. 1907. "The Lower Colorado River and the Salton Basin." *Trans. American Society of Civil Engineers* 59: 1–62.

Grunsky, C. E. 1922. "International and Interstate Aspects of the Colorado River Problem." *Science* 56: 521–27.

Gutiérrez-Galindo, E. A., G. Flores-Muñoz, G. Olguín-Espinoza, M. F. Villa-Andrade, y J. A. Villaescusa-Celaya. 1985. "Insecticidas organoclorados en peces del valle de Mexicali, Baja California, México." *Ciencias Marinas* 14: 1–22.

Gutiérrez-Galindo, E. A., G. Flores-Muñoz, and A. Aguilar. 1988a. "Mercury in Freshwater Fish and Clams from Cerro Prieto Geothermal Field of Baja California, México." *Bulletin of Environmental Contamination and Toxicology* 41: 210–17.

Gutiérrez-Galindo, E. A., G. Flores-Muñoz, and J. A. Villaescusa-Celaya. 1988b. "Chlorinated Hydrocarbons in Molluscs of the Mexicali Valley and Upper Gulf of California." *Ciencias Marinas* 13: 91–113.

Hammond, George P., and Agapito Rey. 1940. *Narratives of the Coronado Expedition, 1540–1542.* Albuquerque: University of New Mexico.

Harding, Benjamin L., Taiye B. Sangoyomi, and Elizabeth A. Payton. 1995. "Impacts of a Severe Drought on Colorado River Water Resources." *Water Resources Bulletin* 31: 815–24.

Hardy, R. W. H. 1829. *Travels in the Interior of Mexico in 1825, 1827, and 1828.* London. Chaps. 13 and 14 and part of chap. 15, pp. 312–84, deal with the stay in the Colorado estuary.

Haro Cordero, Sergio. "Los cucapá asfixiados por políticas gubernamentales." *Sietedías* (Mexicali, México), 20–26 Febrero, 1–4.

Harris, Tom. 1991. *Death in the Marsh.* Washington, D.C.: Island Press.

Harris, Tom, et al. 1992. "David Love" and assorted articles on selenium. *High Country News* 24, no. 2 (10 February).

Hawley, Ellis. 1966. "The Politics of the Mexican Labor Issue, 1950–1965." *Agricultural History* 40 : 157–76.

Hayes, Douglas L. 1991. "The All-American Canal Lining Project: A Catalyst for Rational and Comprehensive Groundwater Management on the United States–Mexico Border." *Natural Resources Journal* 31: 803–808.

Heintzelman, Samuel P. 1857. *Report of July 15, 1853.* 34th Cong., 3rd sess., House Executive Document no. 76. Washington, D.C.: Government Printing Office.

Hendrickson, D.A., and W.L. Minckley. 1984. "Cienegas—Vanishing Climax Communities of the American Southwest." *Desert Plants* 6: 131–75.

Hendrickson, D. A., and A. Varela-Romero. 1989. "Conservation Status of the Desert Pupfish, *Cyprinodon macularius*, in México and Arizona." *Copeia* 2: 478–83.

Hernández-Ayón, J. Martín, M. Salvador Galindo-Bect, Bernardo P. Flores-Báez, and Saúl Alvarez-Borrego. 1993. "Nutrient Concentrations Are High in the Turbid Waters of the Colorado River Delta." *Estuarine, Coastal and Shelf Science* 37: 593–602.

Hinojosa-Huerta, Osvel, Carlos Valdés-Casillas, Francisco Zamora-Arroyo, Yamilett Carrillo-Guerrero, Edward Glenn, Jaqueline García, and Mark Briggs. 1999. "Important Wetland Zones in the Colorado River Delta." Unpublished manuscript.

Hinojosa-Huerta, O., S. DeStefano, and W. Shaw. N.d. "Habitat Use by Yuma Clapper Rails in the Colorado River Delta, Baja California and Sonora, Mexico." Unpublished manuscript.

Hinojosa-Huerta, Osvel, Stephen DeStefano, and William W. Shaw. 2001. "Distribution and Abundance of the Yuma Clapper Rail (*Rallus longirostric yumanensis*) in the Colorado River Delta, Mexico." *Journal of Arid Environments* 49: 171–82.

Holburt, Myron B. 1982. "International Problems on the Colorado River." *Water Supply and Management* 6: 105–14.

———. 1984. "The 1983 High Flows on the Colorado River and Their Aftermath." *Water International* 9: 99–105.

Holsinger, Kent. 2001. "A 7-State Perspective Regarding Management of the Colorado River for the 21st Century." Field Hearing on Colorado River Management, House Resources Committee. July 9, Salt Lake City.

Homer-Dixon, Thomas F. 1991. "Environmental Changes as Causes of Acute Conflict." *International Security* 16 (fall): 76–116.

Horvitz, Steve. 1999. *Salton Sea 101—The Salton Sea, California's Greatest Resource.* North Shore, California: Sea and Desert Interpretive Association.

Hughes, J. D. 1967. *The Story of Man at the Grand Canyon.* Grand Canyon Natural History Association Bulletin 14. N.p.: Grand Canyon Natural History Association.

Humphrey, W. E. 1911, 1912. "Shooting the Vanishing Sheep of the Desert." Pts. 1 and 2. *Outdoor Life* 28 and 29: 477–87, 95–105.

Hundley, Norris, Jr. 1966. *Dividing the Waters: A Century of Controversy between the United States and Mexico.* Berkeley: University of California Press.

———. 1973. "The Politics of Reclamation: California, the Federal Government, and the Origins of the Boulder Canyon Act—A Second Look." *California Historical Quarterly* 52 4 (winter): 292–325.

———. 1975. *Water and the West: The Colorado River Compact and the Politics of Water in the American West.* Berkeley: University of California Press.

———. 1978. "The Dark and Bloody Ground of Indian Water Rights." *Western Historical Quarterly* 9 (): 455–82. Excellent overview of the matter, with notes.

———. *The Great Thirst: Californians and Water, 1770s–1990s.* Berkeley: University of California Press, 1992.

International Boundary and Water Commission (IBWC). "Minute No. 306, Conceptual Framework for the United States–Mexico Studies for Future Recommendations Concerning Riparian and Estuarine Ecology of the Limitrophe Section of the Colorado River and Its Associated Delta." 2 December, International Boundary and Water Commission, United States and Mexico, United States Section, El Paso, Texas.

Israelsen, Brent. 2001. "Groups Want Water Left in Colorado River." The Salt Lake Tribune. March 6, www.sltrib.com/03062001/utah/77140.htm.

Ives, J. C. 1861. *Report upon the Colorado River of the West, explored in 1857 and 1858.* 36th Cong., 1st sess., House Executive Document no. 90. Washington, D.C.: Government Printing Office. Five parts separately paginated and appendices (total 368 pp.). The estuary and Colorado delta up to Yuma are discussed in pt. 1 (general report), pp. 19–44, and pt. 2 (hydrography), pp. 7–10.

James, George Wharton.1906. *The Wonders of the California Desert (Southern California): Its Rivers and Its Mountains, Its Canyons and Its Springs, Its Life and Its History, Pictured and Described.* 2 vols. Boston: Little, Brown.

Jaramillo, A., L. Rojas, and T. Gerrodette. 1999. "A New Abundance for Vaquitas: First Steps for Recovery." *Marine Mammal Science* 15: 957–73.

Johnson, Boma. 1985. *Earth Figures of the Lower Colorado and Gila River Deserts: A Functional Analysis.* Arizona Archeological Society Publication no. 20, *Arizona Archeologist* (November).

Johnson, R. R., and D. A. Jones, eds. 1977. *Importance, Preservation and Management of Riparian Habitats: A Symposium.* General Technical Report RM-43, Rocky Mountain Forest Range Experimental Station, Fort Collins, Colorado.

Jordán, Fernando. 1951. *El Otro México.* México, D.F.: Biografías Gandesa.

Kelly, M. 1993. *Nafta's Environmental Side Agreement: A Review and Analysis.* Austin, Texas: Texas Center for Policy Studies.

Kelly, William H. 1977. *Cocopa Ethnography.* Anthropological Papers of the University of Arizona no. 29, Tucson.

Kennan, George. 1922. *E. H. Harriman.* Boston [publisher not known]

Kerig, Dorothy Pierson. 1988. *Yankee Enclave: The Colorado Land Company and Mexican Agrarian Reform in Baja California, 1902–1944.* Ph.D. diss., University of California, Irvine.

King, K. A. L. Velasco, J. García-Hernandez, B. J. Zaun, J. Record, and J. Wesley. Year. "Contaminants in Potential Prey of the Yuma Clapper Rail: Arizona and California, USA, and Sonora and Baja, California, México, 1998–1999." Contaminants Program Report, U.S. Fish and Wildlife Service, Region 2.

Kino, Eusebio Francisco. 1919. "Celestial favors . . . experienced in the new conquests and new conversions of the new kingdom of New Navarra of this unkown North America, and the land-passage to California in 35 degrees latitude, with the new cosmorgraphic map of these new and extensive lands which hitherto have been unkown . . . ." In *Kino's Historical Memoir of Primería Alta,* 2 vols., trans. H. E. Bolton Cleveland, Ohio: Arthur H. Clark. Father Kino's two journeys of 1701 and 1702 in the Colorado delta are described in vol. 1, bk. 3, chaps. 4–6, pp. 312–22, and vol. 1, bk. 4, chaps. 4–5, pp. 340–46. Spanish text in Publicaciones del Archivo General de la Nación, vol. 8, Secretaría de Gobernación, Mexico City, 1913–1922.

Kittredge, William. 1992. *Hole in the Sky: A Memoir.* New York: Vintage Books.

Kniffen, Fred B. 1929. *The Delta Country of the Colorado.* Ph.D. diss. University of California, What Campus? It is not specified on the text. .

———. 1931. *The Primitive Cultural Landscape of the Colorado Delta.* Lower California Studies III. Berkeley: University of California Publications in Geography 5, 43–66 4.

———. 1932. "The Natural Landscape of the Colorado Delta." In *University of California Publications in Geography* 5, ed. Carl O. Sauer and J. B. Leighly, 149–244. Berkeley: University of California Press.

Kolb, E. L. 1914. *Through the Grand Canyon from Wyoming to Mexico.* New York: Macmillan.

Kowaleski, Michal. 1996. "Taphonomy of a Living Fossil: The Lingulide Brachiopod *Glottidia palmeri* Dall from Baja California, Mexico." *Palaios* 11: 244–65.

Kowaleski, Michal, and Karl W. Flessa. 1995. "Comparative Taphonomy and Fuanal Composition of Shelly Centers from Northeastern Baja California, Mexico." *Ciencias Marinas* 21: 155–77.

———. 1997. "Predatory Scars in the Shells of a Recent Lingulid Brachiopod: Paleontological and Ecological Implications." *Acta Palaeontologica Polonica* 42: 497–532.

Kowaleski, Michal, Glenn A. Goodfriend, and Karl W. Flessa. 1998. "High-Resolution Estimates of Temporal Mixing within Shell Beds: The Evils and Virtues of Time-Averaging." *Paleobiology,* 24 (1998): 2287–304.

Kowaleski, Michal, Guillermo E. Avila Serrano, Karl W. Flessa, and Glenn A. Goodfriend. "A Dead Delta's Former Productivity: Two Trillion Shells at the Mouth of the Colorado River." *Geology,* manuscript in press.

Kresan, Peter. 1997. "A Geologic Tour of the Lower Colorado River Region of Arizona and Sonora."

In *Dry Borders: A Double Issue*, special issue of *Journal of the Southwest* 39 (autumn–winter): 567–612.

Kroeber, Alfred L. [1925] 1976. *Handbook of the Indians of California*. Reprint, New York: Dover.

Kroeber, Clifton B. 1983. *Man, Land, and Water: Mexico's Farmlands Irrigation Policies, 1885–1911*. Berkeley: University of California Press.

Kurtén, Björn, and Elaine Anderson. 1980. *Pleistocene Mammals of North America*. New York: Columbia University Press.

Lao-tse. 1972. *Tao te Ching*. Trans. Gia-fu Feng and Jane English. New York: Publisher.

Lavín, M. F., and Salvador Sánchez. 1999. "On How the Colorado River Affected the Hydrography of the Upper Gulf of California." *Continental Shelf Research* 19: 1545–60.

Leonard, Irving A. 1949. *Books of the Brave: Being Accounts of Books and Men in the Spanish Conquest and Settlement of the Sixteenth Century New World*. Cambridge, Massacusetts: Harvard University Press.

Leon-Portilla, M. 1972. "Descrubrimiento en 1540 y primeras noticias de la isla de Cedros." *Calafia (Revista Universidad Autónoma de Baja California)* 2:8–10.

Leopold, A. S. (Starker). 1950. "Vegetation Zones in Mexico." *Ecology* 31: 507–518.

———. 1959. *Wildlife in Mexico*. Berkeley: University of California Press.

Leopold, Aldo. 1949. *A Sand County Almanac, and Sketches Here and There*. New York: Oxford University Press. See especially "The Green Lagoons," pp. 141–49 for his essay account of the delta journey, written some two decades after the fact.

———. 1953. *Round River*. Ed. Luna B. Leopold. London: Oxford University Press. See especially "The Delta Colorado," pp. 10–30, taken from the New Mexico journals of 14 October–November 1922.

Levy, Marc A. 1995. "Is the Environment a National Security Issue?" *International Security* 20, no. 2 (fall): 35–62.

Lewis, C. S. 1947. *The Abolition of Man*. New York: Publisher. 1947.

Lingenfelter, Richard E. 1978. *Steamboats on the Colorado River 1852–1916*. Tucson: University of Arizona Press.

Luecke, Daniel, Jennifer Pitt, Chelsea Congdon, Edward Glenn, Carlos Valdés-Casillas, and Mark Briggs. 1999. *A Delta Once More: Restoring Riparian and Wetland Habitat in the Colorado River Delta*. Washington, D.C.: Environmental Defense Fund Publications.

Luis Moreno, José, ed. 1992. *Ecología, Recursos Naturales y Medio Ambiente en Sonora*. Obregón, México: El Colegio de Sonora

Lumholtz, Carl. 1912. *New Trails in Mexico*. New York: Charles Scribner's Sons.

MacDougal, D. T. 1904a. "Botanical Exploration of the Southwest." *Journal of the New York Botanical Garden* 5: 89–98. Boat trip from Yuma to Punta San Felipe.

———. 1904b "Delta and Desert Vegetation." *Botanical Gazette* 38: 44–63.

———. 1906. "The Delta of the Rio Colorado." *Bulletin of the American Geographic Society* 38: 1–16.

———. 1907a. "The Desert Basins of the Colorado Delta." *Bulletin of the American Geographic Society* 39 : 705–29. New freshwater fish described.

———. 1908a. "More Changes in the Colorado River." *National Geographic Magazine*, no. 19: 52–54.

———. 1908b. "A Voyage below Sea Level on the Salton Sea." *Outing*: 592–601.

Martínez, María Jesus. 2000. "Humedales del Delta del Río Colorado: Los Humedales de la Reserva de la Biósfera Alto Golfo de California del Río Colorado: Un Esfuerzo Cooperativo." *DUMAC: Ducks Unlimited de México* 22 : 31–34.

Martínez, Oscar J. 1988. *Troublesome Border*. Tucson: University of Arizona Press.

Martínez, Pablo L. [1956] 1991. *Historia de Baja California*. Baja California: Consejo Editorial del Gobierno del Estado de B.C.

Marty, Kevin. 2001. "Nature Plays a Major Role in Cleaning Up Polluted Waters. *Imperial Valley Press*, 7 June.

Massey, B.W., and E. Palacios. 1994. "Avifauna of the Wetlands of Baja California, México." *Studies in Avian Biology* 15: 45–57.

Matthews, J. B. 1969. "Tides in the Gulf of California." In *Environmental Impacts of Brine Effluents on the Gulf of California* U.S. Department of the Interior: Research and Progress Report, no. 387, ed. D. A. Thompson, A. R. Mead, J. R. Shreiber, J. A. Hunter, W. F. Savage, and W.W. Rinne. , 41–50.

McGuire, T., and G. C. Valdez-Gardea. 1997. "Endangered Species and Precarious Lives in the Upper Gulf of California. *Culture and Agriculture* 19: 101–07.

McGuire, T., and J. Greenberg, eds. N.d. *Maritime Community and Biosphere Reserve: Crisis and Response in the Upper Gulf of California*. Bureau of Applied Research in Anthropology, Paper no. 2. Tucson: University of Arizona.

Meine, Curt. 1988. *Aldo Leopold: His Life and Work*. Madison: University of Wisconsin Press.

Mejia, Javier. 1999. "Sólo a Cucupas [sic] Acceso al Delta del Río Colorado." *La Voz de la Frontera* (Mexicali, México), 26 Abril, 21-A.

———. 2000. "Apremian la restauración en el Delta del Colorado." *La Voz de la Frontera* (Mexicali, México), 7 Junio, 1-A, 10-A.

Meldahl, Keith H., and Alan H. Cutler. N.d. "Neotectonics and Taphonomy: Pleistocene Molluscan Shell Accumulations in the Northern Gulf of California." Unpublished manuscript.

Mellink, E. N.d. "On the Wildlife of Wetlands on the Mexican Portion of the Colorado River." *Bulletin of the Southern California Academy of Science*, in press.

———. 1995. "Status of the Muskrat in the Valle de Mexicali and Delta del Río Colorado, México." *California Fish and Game* 81 : 33–38.

Mellink, E., E. Palacios, and S. González. 1996. "Notes on

the Nesting Birds of the Ciénega de Santa Clara Saltflat, Northwestern Sonora, México." *Western Birds* 27: 202–03.

Mellink, E., and J. Luevano. 1998. "Status of Beavers (*Castor canadensis*) in Valle de Mexicali, México." *Bulletin of Southern California Academy of Science* 97: 115–20.

Merriam, R. 1969. "Source of Sand Dunes of Southeastern California and Northwestern Sonora, México." *Geological Society of America Bulletin* 80 531–34.

Metz, Leon C. 1989. *Border: The U.S.–Mexico Line.* El Paso, Texas: Mangan Books.

Meyer, Michael C. 1984. *Water in the Hispanic Southwest: A Social and Legal History, 1550–1850.* Tucson: University of Arizona Press.

Meyers, Charles J., and Richard L. Noble. 1967. "The Colorado River: A Treaty with Mexico." *Stanford Law Review* 19: 367–419.

Miller, Robert Rush, and Lee A. Fuiman. 1987. "Description and Conservation of *Cyprinodon macularius eremus*, A New Subspecies of Pupfish from Organ Pipe Cactus National Monument, Arizona." *COPEIA* 3: 593–609.

Minkley, W. L. 1991. "Native Fishes of the Grand Canyon Region: An Obituary?" In *Colorado River Ecology and Dam Management.* Ed. National Research Council. Place: Publisher. Pages 122 and124 speak about the 45 to 455 million metric tons of silt transported annually through the Grand Canyon between 1922 and 1935.

Minkley, W. L. and James E. Deacon, eds. YEAR. *Battle Against Extinction: Native Fish Management in the American West.* Tucson: University of Arizona Press.

Monks, Vicki. 1992. "Baja's Imperiled Porpoise: The U.S. Taste for an Endangered Mexican Fish Threatens the Extinction of the Elusive Vaquita." *Defenders* 67 (July/August): 10–20.

Mora, M. 1991. "Organichlorines and Breeding Success in Cattle Egrets from the Mexicali Valley, Baja California, México." *Colonial Waterbirds* 14: 127–32.

Mora, M., and D. Anderson. 1991. "Seasonal and Geographical Variation of Organochlorine Residues in Birds from Northwest Mexico." *Archives of Environmental Contamination and Toxicology* 21: 541–48.

———. 1995. "Selenium, Boron, and Heavy Metals in Birds from Mexicali Valley, Baja California, Mexico." *Bulletin of Environmental Contamination and Toxicology* 54: 198–206.

Mora, Miguel A., Jaqueline García, Maria de la Paz Carpio-Obeso, and Kirke King. 1999. "Contaminants without Borders: A Regional Assessment of the Colorado River Delta Ecosystem." Paper presented at the International Congress on Ecosystem Health, Sacramento, California, 15–20 August.

Morris, Milton Berger. 1927. *United States Ambitions on Lower California.* Master's thesis, University of Pittsburgh.

Morrison, Jason I., Sandra L. Postel, and Peter H. Gleick. 1966. *The Sustainable Use of Water in the Lower Colorado River Basin.* A Joint Report of the Pacific Institute for Studies in Development, Environment, and Security and the Global Water Policy Project, Oakland, California: Pacific Institute for Studies in Development, Environment, and Security.

Morrison, Jason, and Michael Cohen. 1999. "Restoring California's Salton Sea." *Borderlines* 7 (January): 1–4.

Moyano Pahissa, Angela. 1983. *California y Sus Relaciones con Baja California: Síntesis del Desarrollo Histórico de California y Sus Repercusiones sobre Baja California.* México, D.F.: Fondo de Cultura Económica.

Mumme, Stephen P. 1996. "The Institutional Framework for Transboundary Inland Water Water [sic] Management in North America: Mexico, Canada, The United States, and Their Binational Strategies." Report submitted to the Commission on Environmental Cooperation, Project 95.23.01, the Mexico–United States–Canada Transboundary Inland Water Project. December.

———. 1999. "Managing Acute Water Scarcity on the U.S.—Mexico Border: Institutional Issues Raised by the 1990s Drought." *Natural Resources Journal* 39 (winter): 149–66.

Mumme, Stephen P., and T. T. Moore. 1999. "Innovation Prospects in U.S.—Mexico Border Water Management: The IBWC and the BECC in Theoretical Perspective." *Environment and Planning* 17: 753–72.

Murphy, Robert Cushman. 1917. "The Desert Life Group, and an Account of the Museum Expedition into Lower California." *Brooklyn Museum Quarterly*, no. 4: 179–210.

Murphy, Robert Cushman. 1917. "Natural History Observations from the Mexican Portion of the Colorado Desert." In *Abstract of Proceedings*, nos. 24–25, Linnaean Society of New York.

Nabhan, Gary. 1978. "Living with a River." *Journal of Arizona History*, no. 19: 1–16.

———. 1979. "The Ecology of Floodwater Farming in Arid Southwestern North America." *Agro-Ecosystems*, no. 5: 245–55.

———. 1982. *The Desert Smells Like Rain: A Naturalist in Papago Country.* New York: Farrar, Straus and Giroux.

———. 1985. *Gathering the Desert.* Tucson: University of Arizona Press.

Nadeau, Remi A. 1950. *The Water Seekers.* Garden City, New York: Doubleday.

Nagler, Pamela, Kara Gillon, Jennifer Pitt, Bill Snape, and Edward Glenn. 2000. "Application of the U.S. Endangered Species Act across International Borders: The Case of the Colorado River Delta, Mexico." *Environmental Science and Policy* 3: 67–72.

Nathanson, Milton. 1980. *Updating the Hoover Dam Documents, 1978.* Washington, D.C.:U. S. Department of the Interior, Bureau of Reclamation.

Neary, John. 1981. "Pickleweed, Palmer's Grass, and Saltwort: Can We Grow Tomorrow's Food with Today's Saltwater?" *Science* 81 (2): 38–43.

Nijhuis, Michelle. 2000a. "Accidental Refuge: Should We Save the Salton Sea?" *High Country News* 32, no. 12 (19 June): 1, 8–13.

———. 2000b. "A River Resurrected: The Colorado River Delta Gets a Second Chance." *High Country News* 32, no. 13 (3 July): 1, 10–13.

Norris, K. S., and W. N. McFarland. 1958. "A New Harbor Porpoise of the Genus *Phocoena* from the Gulf of California." *Journal of Mammals* 39: 22–39.

North, Arthur W. 1910. *Camp and Camino in Lower California.* New York: Baker and Taylor.

Ochoa Zazueta, Jesús Angel. 1973. "Los Cucapá de El Mayor Indígena." *Boletín DEAS*, no. 1: 17–43. A journal of the Instituto Nacional de Antropología e Historia.

Ohmart, Robert D., Wayne O. Deason, and Constance Burke. 1977. "A Riparian Case History: The Colorado River." In *Importance, Preservation and Management of Riparian Habitat: A Symposium*, USDA Forest Service General Technical Report RM-43, ed. R. Roy Johnson and Dale A. Jones., 35–47. Fort Collins, Colorado: Rocky Mountain Forest and Range Experiment Station. Excellent historical comments on cottonwood communities.

Oñate, Juan. 1916. "Journey of Oñate to California by Land (Zárate-Salmerón)." In *Spanish Exploration in the Southwest, 1542–1706*, trans. and ed. Herbert Eugene Bolton. , 268–80. New York: Charles Scribner's Sons.

Orr, David, and Stuart Hill. 1978. "Leviathan, the Open Society, and the Crisis of Ecology." *Western Political Quarterly*, no. 31: 45769.

Oyarzabal-Tamargo, Francisco, and Robert A. Young. 1978. "International External Diseconomies: The Colorado River Salinity Problem in Mexico." *Natural Resources Journal* 18: 77–89.

Page, Gary W., Lynne E. Stenzel, Janet E. Kjelmyr, and W. David Shuford. 1990. "Shorebird Numbers in Wetlands of the Pacific Flyway: A Summary of Spring and Fall Counts in 1988 and 1989." Point Reyes Bird Observatory, February.

Palacios, E., and E. Mellink. 1992. "Breeding Bird Records from Montague Island, Northern Gulf of California." *Western Birds* 23: 41–44.

———. 1993. "Additional Records of Breeding Birds from Montague Island, North Gulf of California." *Western Birds* 24: 259–62.

———. 1996. "Status of the Least Tern in the Gulf of California." *Journal of Field Ornithology* 67: 48–58.

Palacios, E., D. Anderson, E. Mellink, and S. González. 2000. "Distribution and Abundance of Burrowing Owls on the Peninsula and Islands of Baja California." *Western Birds* 31: 89–99.

Pattie, James O. [1833] 1966. *The personal narrative of James O. Pattie of Kentucky . . . .* Ed. Timothy Flint. Cincinnati, Ohio: E. D. Flint. Reprinted, Ann Arbor, Michigan: University Microfilms.

Peña Ceceña, Erasmo. 2000a. "Respeto a sus Derechos Demandan Cucapás." *La Voz de la Frontera* (Mexicali, México), 4 Marzo 2000, 3A, 10A.

———. 2000b. "Excedentes del Río ya son Sólo Historia." *La Voz de la Frontera* (Mexicali, México), 23 Septiembre.

Peresbarbosa, E., and E. Mellink. 1994. "More Records of Breeding Birds from Montague Island, Northern Gulf of California." *Western Birds* 25: 201–2.

Perez U., Matilde. 2001. "Exige Dassur revisar la distribución de agua en el delta del río Colorado." www.jornada.unam.mx/2001/may01/010525/048n2soc.html

Piest, Linden, and José Campoy. 1998. "Report on the Yuma Clapper Rail Surveys at Ciénega de Santa Clara, Sonora." Arizona Fish and Game Department, Region 4, Place.

Pineda Pablos, Nicolás, ed. 1998. *Hermosillo y el Agua: Infraestructura Hidráulica, Servicios Urbanos y Desarrollo Sostenible.* Hermosillo, México: El Colegio de Sonora.

*El Pionero* (Mexicali). 2000. "Advierten a CILA problemas por manejo del Río Colorado." 6 Octubre, 1A.

Pitt, Jennifer. 2001. "Can We Restore the Colorado River Delta?" *Journal of Arid Environments*, no. 49: 211–20.

Pitt, Jennifer, Daniel F. Luecke, Michael Cohen, Edward P. Glenn, and Carlos Valdés-Casillas. 2000. "Two Countries, One River: Managing for Nature in the Colorado River Delta." *Natural Resources Journal*, no. 40 (fall): 819–64.

Poff, LeRoy N., J. David Allen, Mark B. Bain, James R. Karr, Karen L. Prestegaard, Brian D. Richter, Richard E. Sparks, and Julie C. Stromberg. 1997. "The Natural Flow Regime." *Bioscience* 47: 769–84.

Pontius, Dale. 1997. *Colorado River Basin Study: Report to the Western Water Policy Review Advisory Commission*, 15. Place: Publisher.

Postel, Sandra. 1992. *Last Oasis: Facing Water Scarcity.* New York: W. W. Norton.

———. 1999. *Pillar of Sand: Can the Irrigation Miracle Last?* New York: W. W. Norton.

———. 2000. "Troubled Waters." *Utne Reader* (1 July): 62–66.

———. 2001. "Growing More Food with Less Water." *Scientific American* 284 (February): 46–51.

Powell, John Wesley. 1891. "New Lake in the Desert." *Scribner's Magazine* 10 : 463–8.

———. [1875] 1987. *The Exploration of the Colorado River and Its Canyons.* Reprint, New York: Penguin.

Presser, Theresa S., Marc A. Sylvester, and Walton H. Low. 1994. "Bioaccumulation of Selenium from Natural Geologic Sources in Western States and Its Potential Consequences." *Environmental Management* 18: 423–36.

Pryde, Philip R. 2001. "A Layperson's Guide to the Water Transfer Agreements: What the 'Babbitt Initiatives' Mean for Southern California." *San Diego Earth Times*, June. http://www.sdearthtimes.com/et0601/et0601s2.html.

Reisner, Marc. [1986] 1993. *Cadillac Desert: The American West and Its Disappearing Water.* Reprint, New York: Penguin.

*Report of the American Section of the International Water Commission United States and Mexico.* 1930. Referred to the Committee on Foreign Affairs. Washington, D.C.: Government Printing Office.

Reyes, Carlos. 2001. "Urgen México y EU aclarar uso del agua." *El Norte*. 23 Febrero. www.elnorte.com//nacional/articulo/092521

Richardson, Jim, and Jim Carrier. 1992. *The Colorado: A River at Risk*. Englewood, Colorado: Westcliffe Publishers.

Rocke, T. E., and M. Friend. 1999. "Wildlife Disease in the Colorado Delta as an Indicator for Ecosystem Health." Paper presented at the Third International Congress on Ecosystem Health, Sacramento, California, 15–20 August. Rockwood, Charles R. [1909] 1930. *Born of the Desert*. Reprint, Calexico, California: *Calexico Chronicle*.

Rodríquez Moreno, Carina. 2000. "Prohíben pesca a Cucupás." *La Crónica* (Mexicali, México), 28 Febrero: 1A.

———. Year. "Salen a la Pesca: Desafían Cucapás a militares." *La Crónica* (Mexicali, México), day Month, 1A.

Rodríquez Moreno, Carina, y Nerelda Romero. 2000. "Retan cucapás a Semarnap." *La Crónica* (Mexicali, México), 29 Marzo, 1A.

Rodriquez, C. A., K. W. Flessa, M. A. Tellez-Duarte, D. L. Dettman, and G. A. Avila-Serrano. 2001. "Macrofaunal and Isotopic Estimates of the Former Extent of the Colorado River Estuary, Upper Gulf of California, Mexico." *Journal of Arid Environments*, no. 49: 183–94.

Rodriquez, Carlie A., Karl W. Flessa and David L. Dettman. "Effects of Upstream Diversion of Colorado River Water on the Estuarine Bivalve Mollusc *Mulina Coloradoensis*." *Conservation Biology*, ms. in press.

Rogers, Peter. 1993. "The Value of Cooperation in Resolving International River Basin Disputes." *Natural Resources Forum* (May): 117–31.

Rojas, L., and B. Taylor. 1999. "Risk Factors Affecting the Vaquita *(Phocoena sinus)*." *Marine Mammal Science* 15: 974–98.

Rojas-Bracho, L., and J. Urbán-Ramírez. "Vaquita: Its Environment, Biology and Problematic." In *Coastal Management in Mexico: The Baja California Experience*, ed. J. L. Fermán-Almada, L. Gómez-Morin, and D. Fischer. New York: American Society of Civil Engineers.

Román, M. 1998. "Managing Wetlands in Northwestern Mexico." *Endangered Species Bulletin* 23: 12–13.

Román, M., and G. Hammann. 1997. "Age and Growth of Totoaba *(Totoaba macdonaldi)* in the Upper Gulf of California." *Fishery Bulletin* 95: 620–28.

Romer, Margaret. 1922. *A History of Calexico*. Annual publication of the Historical Society of Southern California.

Rosenberg, Kenneth V., Robert D. Ohmart, William C. Hunter, and Bertin W. Anderson. 1991. *The Birds of the Lower Colorado River Valley*. Tucson: University of Arizona Press.

Rosenblum, Mort. 2001. "Special Report: Shrinking Water Reserves, Fighting over the Last Drops at the Colorado River's End." *The Desert Sun* (Palm Springs, California), 21 May.

Ruiz-Campos, G., y M. Rodríquez-Meraz. 1997. "Composición Taxonómica y Ecológica de la Avifauna de los Ríos El Mayor, Hardy y Areas Adyacentes, en el Valle de Mexicali, Baja California, México." *Anales del Instituto de Biología*, UNAM (Serie Zoología), no. 68: 291–315.

Russell, Israel. 1898. *Rivers of North America*. New York.

Russell, Stephen M., and Gale Monson. 1998. *The Birds of Sonora*. Tucson: University of Arizona Press.

SAIC/Jones and Stokes. 2002. Second administrative draft. Conservation plan for the Lower Colorado River Multi-Species Conservation Program. (JS00-450.) January 25. Sacramento, California. Prepared for LCR MSCP Steering Committee, Santa Barbara, CA.

Sale, Kirkpatrick. 1985. *Dwellers in the Land: The Bioregional Vision*. San Francisco: Sierra Club Books.

Sánchez Zambrano, Eneida. 2000. "Denuncian cucapás la actuación de mala fe." *La Crónica* (Mexicali, México), 11 Marzo, 15A.

Schoenherr, Allan A. 1988. "A Review of the Life History and Status of the Desert Pupfish, *Cyprinodon macularius*." *Bulletin of the Southern California Academy of Sciences* 87: 104–34.

Schonfield, Robert. 1968. "The Early Development of California's Imperial Valley. *Southern California Quarterly*, no. 50: 279–307, 395–426.

Secretaría de Medio Ambiente, Recursos Naturales y Pesca (SEMARNAP). 1995. *Programa de Manejo Reserva de la Biósfera Alto Golfo de California de California y Delta del Río Colorado*. Publicación Especial, Serie Areas Naturales Protegidas 1. México, D.F.

Shaw, Christopher A. 1981. *The Middle Pleistocene El Golfo Local Fauna from Northwestern Sonora, Mexico*. Mater's thesis, California State University, Long Beach.

Sherbrooke, Wade C. N.d. *Horned Lizards: Unique Reptiles of Western North America*. N.p.: Southwest Parks and Monuments Association.

Sheridan, Thomas E. 1998. "The Big Canal: The Political Ecology of the Central Arizona Project." In *Water, Culture, Power: Local Struggles in a Global Context*, ed. John M. Donahue and Barbara Rose Johnston, 163–86. Washington, D.C.: Island Press.

Siber, G. K. 1990. "Occurrence and Distribution of the Vaquita *Phocoena sinus* in the Northern Gulf of California." *Fishery Bulletin* 88: 339–46.

Simon, Darren. 2001. "Water Transfer Hinges on Habitat Conservation Plan." *Imperial Valley Press*. 15 March.

Simonian, Lane. 1995. *Defending the Land of the Jaguar: A History of Conservation in Mexico*. Austin: University of Texas Press.

Singlemann, Peter. 1978. "Rural Collectivization and Dependent Capitalism: The Mexican Collective Ejido." *Latin American Perspectives* 5 (summer): 38–61.

Smith, Henry Nash. 1950. *Virgin Land: The American West as Symbol and Myth*. Cambridge, Massachusetts: Harvard University Press.

Smythe, William E. 1900. "An International Wedding." *Sunset* 5 (October): 286–300.

———. [1899, 1905] 1969. *The Conquest of Arid America*. Reprint, Seattle: University of Washington Press.

Snape, William. 1998. "Adding an Environmental Minute to the 1944 Water Treaty: Impossible or Inevitable?" Appendix D in *Workshop Proceedings, Water and Environmental Issues of the Colorado River Border Region*, San Luís Río Colorado, Sonora, México, April 30. Defenders of Wildlife and the Pacific Institute for Studies in Development, Environment, and Security.

Soucie, Gary A. 1998. "Research into the Publication Options on the Colorado River and the Northern Gulf of California, including Other Transboundary Water Issues." Report to Defenders of Wildlife, 16 July.

Southwest Center for Environmental Research and Policy. 1999. "The U.S.–Mexican Border Environment: A Road Map to a Sustainable 2020." Border Environment Research Reports. San Diego: Southwest Center for Environmental Research and Policy. Report on Border Institute I. Rio Rico, Arizaon, 7–9 December 1998.

Spamer, E. E. Year. *Bibliography of the Grand Canyon and Lower Colorado River from 1540*. Grand Canyon, Arizona: Grand Canyon Natural History Association.

Stanford, J. A., and J. V. Ward. 1986. "The Colorado River System." In *The Ecology of River Systems* ed. B. R. Davies and K. F. Walker, 353–74. Dordrecht, Netherlands: Publisher.

Starr, Kevin. 1973. *Americans and the California Dream, 1850–1915*. New York: Oxford University Press.

Stegner, Wallace. 1992 *Beyond the Hundredth Meridian: John Wesley Powell and the Second Opening of the West*. New York: Penguin.

Stone, Connie L. 1991. *The Linear Oasis: Managing Cultural Resources Along the Lower Colorado River*. Cultural Resources Series no. 6. Phoenix: Bureau of Land Management.

Strand, Carl L. 1981. "Mud Volcanoes, Faults, and Earthquakes of the Colorado Delta Before the Twentieth Century." *San Diego Journal of History* 27: 43–63.

Stromberg, J. C. 2001. "Restoration of Riparian Vegetation in the Southwestern United States: Importance of Flow Regimes and Fluvial Dynamism." *Journal of Arid Environments*, no. 49: 17–34.

Sykes, Godfrey. 1915. "The Isles of California." *Bulletin of the American Geographical Society* 47: 745–61.

———. 1938. "End of a Great Delta." *Pan-Pacific Geologist*, no. 69: 241–48.

———. 1937. *The Colorado Delta*. American Geographical Society Special Publication no. 19. Carnegie Institution of Washington and American Geographical Society of New York.

———. [1945] 1984. *A Westerly Trend*. Reprint, Tucson: Arizona Historical Society and University of Arizona Press.

Tapia Landeros, Alberto. 1998. *En el Reino de Calafia*. Mexicali, México: Universidad Autónoma de Baja California

Thompson, Robert Wayne 1967. "Tidal Currents and General Circulation." In *Environmental Impact of Brine Effluents on the Gulf of California*, U.S. Dept. Int. Res. Dev. Prog. Rep., ed. D. A. Thompson, A. R. Mead, J. R. Shreiber, J. A. Hunter, W. F. Savage, and W.W. Rinne. , 41–50. Place: Publisher.

Tobin, Mitch. 2001. "Colorado River Is Farmer's Lifeblood," "Accidental Marsh Unveils Past," "Tribe Suffers with Delta," "Mudflats Are Gulf's Nursery." *Arizona Daily Star*, 4 June. www.azstarnet.com. Four-part series on delta of the Colorado River.

Torres Nachon, Claudio. 2000. Denuncia popular de la Ley General del Equilibrio y la Protección al Ambiente ante Procudaría Federal de Protección al Ambiente. El Centro de Derecho Ambiental e Integración Económica del Sur (DASSUR). 23 Junio, Xalapa, México.

———. 2001. Queja ante Comisión Nacional del los Derechos Humanos sobre la Presunta Violación a los Derechos Humanos de los Indígenas Cucapá. El Centro de Derecho Ambiental e Integración Económica del Sur (DASSUR). 30 Junio, Xalapa, México.

Tout, Otis B. N.d. *The First Thirty Years, 1901–1931: Being an Account of the Principal Events in the History of the Imperial Valley, Southern California, U.S.A.* San Diego: Otis B. Tout Publisher.

Trava Manzanilla, José Luis, Jesus Román Calleros, y Francisco A. Bernal Rodríguez, comps. 1991. *Manejo ambientalmente adecuado del agua en la frontera México–Estados Unidos*. Mexicali, México: El Colegio de la Frontera Norte.

de Ulloa, Francisco. [N.d.] 1904. *A Relation of the discovery, which . . . the Fleete of the right noble Fernando Cortez . . . made, . . . out of which Fleete was Captaine the right worshipfull knight Francis de Ulloa . . . Taken out of the third volume of the voyages gathered by M. John Baptista Ramusio*. In "Principal Navigations," Hakluyt Society Publications, Extra Series, vol. 9, Glasgow, 206–278. Reprint in New York: E. P. Dutton. See pp. 214–15 for passage dealing with head of the Gulf of California.

U.S. Fish and Wildlife Service. *Desert Pupfish Recovery Plan*. Albuquerque, New Mexico: U.S. Fish and Wildlife Service.

U.S. Newswire. 1997. "New River Named One of Nation's Most Threatened Rivers." 16 April.

Unitt, P. 1987. "*Empidomax traillii extimus*: An Endangered Subspecies." *Western Birds* 18: 137–62.

Valdés-Casillas, Carlos, Edward Glenn, Mark Briggs, Christopher Lee, Chelsea Congdon, Don Baumgartner, Yamilett Cassillo-Guerrero, Elena Chavattia-Correa, Jaqueline García, Osvel Hinojosa-Huerta, Pat Johnson, Judy King, Dan Luecke, Manuel Mñoz-Viveros, and Jim Riley. N.d. "Revegetation of the Colorado River Delta, Mexico, by Waste Waters from the United States." Unpublished ms.

Valdés-Casillas, Carlos, E. P. Glenn, O. Hinojosa-Huerta, Y. Carrillo, J. García-Hernández, F. Zamora-Arroyo, M. Muñoz-Viveros, M. Briggs, C. Lee, E. Chavarria-

Correa, J. Riley, D. Baumgartner, and C. Congdon. 1998. *Wetland Management and Restoration in the Colorado River Delta: The First Steps.* México, D.F.: CECARENA-ITESM Campus Guaymas and NAWCC.

Valdés-Casillas, Carlos, Yamilett Carrillo Guerrero, Osvel Hinojosa-Huerta and Edward P. Glenn. 1999. "The Colorado River Delta." *Pronatura*, no. 6 (summer): 14–21.

Valdés-Casillas, Carlos, Osvel Hinojosa-Huerta, Framcisco Zamora-Arroyo, Yamilett Carrillo-Guerrero, Carlos Padilla, María López-Camacho, Edward P. Glenn, Jaqueline García-Hernandez, Pamela Nagler, Dean Radtke, Hugo Rodríguez, Elena Chavarrían Correa, Jose Luis Blanco, William Shaw, Mark Briggs, Steve Cornelius and Stephen DeStefano. 2000. "Participatory Habitat Management and Monitoring of Migratory Waterbirds and Associated Wildlife in the Colorado River Delta: A Binational Joint Venture." Interim report presented to North American Wetlands Council. July.

Van Andel, Tjeerd H., and George G. Shor, Jr., eds. 1964. *Marine Geology of the Gulf of California.* Memoir 3. Tulsa, Oklahoma: The American Association of Petroleum Geologists.

Van Dyke, John C. [1903] 1980. *The Desert.* Reprint, Park City, Utah: Peregrine Smith.

Varady, Robert G., Katherine B. Hankins, Andrea Kaus, Emily Young, and Rober Merideth. 2001. ". . . to the Sea of Cortés: Nature, Water, Culture in the Lower Colorado River Basin and Delta—an Overview of Issues, Policies, and Approaches to Environmental Restoration." *Journal of Arid Environments*, no. 49: 195–209.

Velazquez Fierro, Oscar. 1998. "Se Quejan Grupos Etnicos del Estado de Marginación." *La Voz de la Frontera* (Mexicali, México), 16 Diciembre, 3-A.

Velush, Lukas. 2001. "State Water Plans at Odds with Salton Sea Salvage Efforts." *The Desert Sun* (Palm Springs, California), 7 June.

Vidal, O. 1990. "Population Biology and Exploitation of the Vaquita, Phocoena sinus." International Whaling Commission, Science Document SC/42/SM22.30.

Vincent, Kathryn. 1999. "Science and Policy in 'The Hollow of God's Hand.'" *UC MEXUS News*, no. 37 (summer): 1, 4–7.

Wagner, Henry R. 1927. "Some Imaginary California Geography." *Proceedings of the American Antiquarian Society*, 83–129. Worcester, Massacusetts.

Ward, Evan. 1999. "Two Rivers, Two Nations, One History: The Transformation of the Colorado River Delta since 1940." *Frontera Norte*, no. 11 (Julio–Diciembre): 113–40.

*Water Issues in the Colorado River Basin Border Region.* 1999. Workshop Proceedings, Mexicali, 18–19 November.

Oakland, California: Pacific Institute for the Studies in Development, Environment, and Security.

Waters, Frank. [1902] 1974. *The Colorado.* New York: Holt, Rinehart and Winston.

Weidensaul, Scott. 1999. *Living on the Wind: Across the Hemisphere with Migratory Birds.* New York: North Point Press.

Weisman, A. N.d. "Delta of the Colorado River—Past and Present: A Cultural and Historical Overview." Unpublished ms.

Wells, R. S., B. G. Würsig, and K. S. Norris. Year. "A Survey of Marine Mammals of the Upper Gulf of California, México, with an Assessment of the Status of *Phocoena sinus*." Final Report to U.S. Marine Mammals Commission in Fulfillment of Contract MM1300958-0.

Wharton, James George. 1911. *The Wonders of the Colorado Desert.* Boston: Little, Brown.

Wilbur, Ray L., and Northcutt Ely. 1948. *The Hoover Dam Documents A17.* N.p.

Wilbur, Sanford R. 1987. *Birds of Baja California.* Berkeley: University of California Press.

Wilson, Edward O. 1994. *The Diversity of Life.* London and New York: Penguin.

Woodward, Arthur. 1966. *The Republic of Lower California, 1853–1854.* Los Angeles: Dawson's Bookshop.

Worster, Donald. 1985. *Rivers of Empire: Water, Aridity, and the Growth of the American West.* New York: Pantheon.

———. 1984. "History as Natural History." *Pacific Historical Review*, 53: 1–19.

Wright, Harold Bell. 191. *The Winning of Barbara Worth.* Chicago: Book Supply Company.

Zamora, Emilio Lopez. 1977. *El Agua, La Tierra: Los Hombres de México.* México, D.F.: Fondo de Cultura Económica.

Zamora-Arroyo, F., P. Nagler, M. Briggs, D. Radke, H. Rodriques, J. Garcia-Hernandez, C. Valdés-Casillas, A. Huete, and E. P. Glenn. 2001. "Regeneration of Native Trees in Response to Flood Releases from the United States into the Delta of the Colorado River, Mexico." *Journal of Arid Environments*, no. 49: 49–64.

Zazueta, Quintero Carlos. 1978. *La Formación de la Frontera Norte: El Caso de Baja California Norte.* Master's thesis, Centro de Estudios Internacionales de El Colegio de México, México, D.F.

Zengel, S., and E. P. Glenn. 1996. "Presence of the Endangered Desert Pupfish (*Cyprinodon macularius, Cyprinodontidae*) in the Ciénega de Santa Clara, México, Following an Extensive Marsh Dry-Down." *Southwestern Naturalist* 41: 73–8.

Zengel, S. A., V. J. Meresky, E. P. Glenn, R. S. Felger, and D. Ortiz. 1995. "Ciénega de Santa Clara, a Remnant Wetland in the Río Colorado Delta (Mexico): Vegetation Distribution and the Effects of Water Flow Reduction." *Ecological Engineering* 4: 19–36.

# Index

# About the Author

CHARLES BERGMAN began writing about endangered animals, endangered places, and wildlife in general when he was a graduate student at the University of Minnesota. While studying for his Ph.D. in the poetry of the English Renaissance—the age of Shakespeare—he began birding with excellent birders in the state. He discovered that he loved being in the company of wild creatures as much as he loved any of Shakespeare's beautiful sonnets. Soon he combined his love of birds and other animals with his love of writing, beginning to write about his experiences with such creatures as boreal owls in Minnesota and spotted owls in Washington.

Nothing makes him happier than to be in a remote landscape in the company of wild animals. As he began his career as a professor of English at Pacific Lutheran University, he continued to pursue his two passions—nature and writing. He began to teach the literature of nature to college students, often taking them to such distant locations as northern Minnesota, rural England, and Queensland in Australia to read about nature on site. (He has won his university's Faculty Excellence Award for his teaching and writing.) His writing and photography became the vehicles through which he has been privileged to observe and study many of the rarest animals in the world, from wolves on Ellesmere Island to Iberian lynxes in Andalucîa, Spain.

He writes regularly for national magazines such as *Smithsonian, National Geographic, Orion, Natural History, Wildlife Conservation, Defenders, National Wildlife, Audubon*, and many others. His two earlier books focus on the human relationship with nature and animals, particularly the way culture shapes our understanding of nature, and the way nature informs human values. They are, *Wild Echoes: Encounters with the Most Endangered Animals in North America* and *Orion's Legacy: A Cultural History of Man as Hunter.*

While serving undercover for the U.S. Fish and Wildlife Service for an *Audubon* story on parrot smuggling across the Río Grande River in Texas, Bergman realized that he needed to know more about this vast culture to the south—and its wilderness riches. Thus began his love affair with the Spanish language and with Hispanic cultures—both peninsular and American. He has taught at the Universidad de Oviedo in northern Spain and been a Fulbright Fellow in Mexico at Universidad La Salle, where he began his research on the delta of the Colorado River. He now loves especially to write about wildlife and nature in South and Central America. His latest passion in poetry is Pablo Neruda. Few joys can compare with reading the great Nobel Prize winner, a poet of love and nature, in the original Spanish.

Bergman lives with his wife, Susan, on the shores of Puget Sound in the oldest town in Washington State—Steilacoom. They treasure their views of the Sound and the distant Olympic Mountains. They also share their wooded yard with about 65 species of birds that visit their bird gardens, as well as deer, raccoons, opossums, coyotes, foxes, and cottontail rabbits. He has two sons, both in college. Their three cats are heavily belled and join him every day as he writes and reads.